Practical Tideway Construction

for civil engineering contractors
and engineers

Jack Preston

with a foreword by
Sir Edgar Beck, C.B.E.
President · John Mowlem and Co., P.L.C.

Published by Attic Books
The Folly, Rhosgoch, Painscastle, Builth Wells, Powys LD2 3JY
First published 1989

British Library in Publication Data
Preston, Jack, 1920–
Practical tideway construction: a manual
for civil engineering contractors and
engineers
1. Channels
I. Title
627
ISBN 0-948083-14-X

Printed in England by J. W. Arrowsmith Ltd, Bristol

Foreword

I have read Jack Preston's manuscript of "Practical Tideway Construction" with pleasure. It is surprising that such a store of practical wisdom within the ranks of civil engineers experienced on works in tidal waters has not been published before in an informative and comprehensive form.

Much of the text recalls my own early experience with Mowlem in the 1930s, when I was first involved with marine work at Southampton and at Dover. We then had contracts for the King George V Graving Dock for the "Queen Mary" at Southampton and the Southern Railway train ferry dock at Dover. Since then, as Agent, Director and Chairman until 1979, I have keenly watched and appreciated the struggles of the many engineers in the firm who, like Jack Preston, have performed valiant service on all kinds of construction along the rivers and shores of Great Britain and so many places abroad.

The book is full of sound advice, enlivened by anecdotes of success and failure in that most fickle and demanding environment of tide, wind and water. Nothing could better demonstrate the variety and excitement of civil engineering as a career. The text has a refreshing directness and a minimum of jargon in spite of the technicality of the subject. It should encourage all young men to become civil engineers – an outdoor life of great interest together with tremendous satisfaction in being able to see the fruits of one's labours.

The author is not afraid to express his opinions, which are based on 50 years' working life on and around the Thames, Medway, English Channel and other tideways. Young engineers would do well to consider the time-honoured solutions to civil engineering problems which he describes. These may prove better and cheaper than modern technology in many circumstances.

I commend this book to any civil engineer who is to work on, over or under water.

Edgar Beck

London
1988

Contents

5 Site works

Author's introduction

The tidal reaches of rivers in the United Kingdom form waterways which present similar problems to the construction industry worldwide, in spite of their local and particular idiosyncrasies. This book sets out some practical aspects of marine construction which, I believe, will be of great use to all engaged on projects in tidal rivers.

My experience of tidal construction is widespread and extends, not only into rivers, but includes open harbours and sea defences. However, most of this experience was based on the Thames and Medway over a period of half a century. It presented little extra challenge, apart from coping with storms, to carrying out work in harbours and along coastlines and rivers elsewhere.

Work over water, particularly in tideways, calls for an attitude of mind needing far more thought and comradeship than ashore. There is no truer saying than that time and tide wait for no man! More planning is required than ashore in order that no opportunity is missed to squeeze a little more out of the day, before an adverse tide puts paid to it. In addition careful protection of finished work before the tide rises is vital.

When afloat on a working craft, however near or far from shore, there is a sense of remoteness from the mainland requiring those on board to work as a team, wholly reliant on each other. I gained respect for the crew which carries out its duties in conditions far more arduous than those ashore, demanding extra skill and self-assurance. This outlook becomes a sixth sense amongst all those working in a tideway, whatever their job.

After long periods of association with construction in tidal conditions, construction ashore comes more easily. The area of work does not roll about or strive to get away from its moorings and plant stays where left without having to be lashed down. Drive a peg in the ground and it will stay there; apart from a driven pile, the stationary peg in water has yet to be invented. The theodolite or 'dumpy' is no use on board either, so setting out needs some thought. It's a different world afloat!

Jack Preston November 1988
Meopham Green, Kent.

Acknowledgements

The author wishes to record his grateful thanks to the following for their unstinted help and encouragement in the preparation of this book.

The Directors and staff of John Mowlem and Company PLC

The librarian and staff of The Institution of Civil Engineer's library

The Managing Director and Directors of John Shelbourne and Company Ltd. for permission to reproduce the photograph of the berthing of HMS Belfast at Tower Bridge, London.

P. T. Sanderson Esq.

My wife Mary.

1. Introduction

Construction in tidal waters has been practised for thousands of years. The Romans drove foundation piles from a barge. Whereas now we have sophisticated equipment in the construction industry, the basic principles do not seem to have advanced that much. Originally cranage consisted of a timber "crane" jib, with guy ropes to retain the required angle and radius, and a treadmill alongside as a power pack to turn the joist drum. Pile frames were of much the same design as today and there is evidence that treadmills, for this purpose, were still in use in 1750. The present day method of retaining the edge of a berth in a river is called campsheeting – a Roman word! I have been told, on good authority, that the Romans built the original clay river walls in the South East of England to reclaim land.

The length of the tideway in the Thames was shortened in the 14th Century by introducing "staunchers" or sluices, thus making the upper reaches more navigable. Twenty three such staunchers were built in the 62 miles between Oxford and Maidenhead. Locks, as known today, were first introduced on the Thames in the 18th Century.

Although the Romans first built lighthouses ashore, timber and later stone ones were erected in tidal conditions from the 18th Century. A century before, dockyards were constructed in dressed granite, with foundations, sluices and culverts well below low water level, requiring considerable knowledge of cofferdams to enable the work to be carried out. Interlocking timber piles had already been in use in the late 16th Century. The 17th Century also saw the commencement of the development of the Cinque Ports and the massive construction work they involved. A breakwater at Ramsgate, Kent, was one of the earliest to be built: here, caissons were used as a basis of design. In Cherbourg an attempt was made to use caissons about 22 metres high, 45 metres in diameter at the base tapering to 18 metres diameter at the top. These were floated by some 1200 cubic metres of timber and towed to their positions by rowing boats, before being filled with stone. Eighteen out of twenty three were built before disaster struck in a storm and the scheme abandoned.

The growth of industrial development brought about a greater use of tidal rivers, requiring wharves and jetties. At the same time the original railway companies built their own ports around the coasts of the U.K. which were used a great deal as an outlet for coal. Dredging for bridge foundations was introduced into England in 1738.

There are drawings in existence showing a common way of constructing a wharf front in the 19th and early 20th Century. Remembering that sailing craft were almost universal at this time, it was quite adequate to sink a disused barge on the foreshore and fill it with ballast, etc., to form a platform to work from and for a small craft to come alongside. This type of structure was often extended, later on, by timber frontages and in a few cases became extensive wharves for shipping.

Timber jetties and dolphins (see Page 46) were also adequate to accommodate the largest of merchantmen when a "10,000 tonner" was any captain's ultimate goal. I had the nerve-wracking experience in the 1950s of ensuring that there would be sufficient water at low water for the largest oil tanker then afloat, to lie alongside a jetty for several days. This was the ship's maiden voyage and was all of 30,000 tons!

At the time of my introduction to marine work in the mid-1930s steam power for cranes and piling hammers was usual. Handmade timber pile frames, up to 17 metres high, were giving way to steel, although my Company was still using timber in the late 1960s.

Lying in one corner of the Depot was an old "Ringing Engine". This consisted of a short timber pile frame with a main hoist rope leading over the cathead to a small drop hammer. This rope was connected to about five other short ropes at its other end so that a gang of men could act in unison to pull on the hoist rope. This operation took the form of a set of bell ringers, to lift and drop the hammer to drive a pile. Main lifting was carried out by the same method.

From soon after the Second World War, reinforced concrete piles were used extensively for foundation work and these coincided with the increase in the construction of reinforced concrete jetties.

Timber piled and planked wharf fronts, which had often been driven in advance of other defective fronts, were covered by steel sheet piling and existing ties extended or new ones introduced. In more recent years, when it seemed everything had to be longer or stronger, ground anchors became the vogue.

Pre-stressed concrete had a brief spell with the advantage of construction in bolted panels and smaller section piles. It was soon out of favour once it was discovered that the only practical method of repair was in conventional reinforced concrete.

Diesel soon followed petrol for motive power. "Delmag" piling hammers were used a great deal and other makes followed. These had the capacity to drive steel box piles and modern large diameter tubular piles – far beyond the capabilities of the "drop" and old steam hammers. Vibrating hammers were also developed for driving into hard clay and ballast.

As ships have become larger the strength of marine structures and their fendering has increased to withstand the berthing and mooring loads but if a large ship is determined on destruction no conventional structure can withstand the forces applied by it.

Floating craft for marine construction have increased in size and performance since the 1950s. At that time a steel or sailing barge was stripped down and heavy ballast placed in the bottom of the hold. A 3 tonne ballast waggon was then set on rails athwart in order to counteract lifting of loads overside. Once the decking was fixed a 17 metre timber pile frame was erected with a 2–3 tonne piling winch within its base. This "power pack" could handle a drop hammer of equivalent weight. Four 1 tonne hand operated crab winches and a 1½ tonne winch fore and aft, all connected to barge anchors, were the means of stabilising a position or controlling movement on site.

This type of craft does not compare with the modern pontoons and diesel driven deck winches and multi-tonne anchors.

At the risk of commending "the good old days", I recall a normal one and a half day sequence for setting up a position to drive concrete piles ashore:-

Day 1	7.00 a.m.	Five men load timber pile frame on to 5 ton lorry by hand – drive to site with two men and driver in cab and the other three amongst the frame on the back. Unload and erect frame horizontally while lorry returns for piling winch and hammer. Unload winch behind base of frame and hammer as "dead man" behind the winch. Insert 6 m × 300 × 150 timber "kickers"
	5.00 p.m.	between winch and frame, and rear frame to upright position
Day 2	7.00 a.m.	Fit winch into frame. Bring hammer into leaders. Jack up frame to insert timber running road and rollers under. Move frame by winch power to position of first pile and fix three guy wires from cat-head.
	1.00 p.m.	First piles delivered to site and unloaded, using pile frame and winch. A crane was unheard of for any lifting operations.
	4.50 p.m.	Telephone the office to report number of piles driven, penetration, etc., and order more piles.

Apart from the one trip on the lorry the men travelled by public transport to home and site.

I never saw such a quick set up again, once the timber frame was discarded. Factory Inspectors refused to allow further new frames to be built and used!

The development of statutory governing authorities

Before 1939 saw numerous Land Drainage Commissioners and similar institutions responsible for land drainage, flood defence and usually about one river in each County. Some made charges in the form of a levy or "Scot" to finance the work they carried out and to augment State Grants.

Under the River Board Act 1948, these small bodies were grouped together to form River Boards in each county. The Water Resources Act 1963 formed new bodies known as River Authorities with more powers and the Water Act 1973 was responsible for the formation of the present ten Water Authorities (see Page 66). These are responsible for flood prevention and pollution (other than that by oil where Port and Harbour Authorities prevail). Their other responsibilities are not relevant to this text. Harbour and Port Authorities are concerned with the prevention of oil pollution.

Similar authorities are to be found throughout the world and, in many instances, will have similar requirements to statutory authorities in the United Kingdom.

Regulations

In the present day there is generally a governing Authority for the tidal reaches of every river and for each port or harbour, both in the U.K. and overseas. For example, the Port of London Authority covers the River Thames and acts as agent for those areas owned by the Duchy of Cornwall; the Southern Water Authority the upper reaches of the River Medway, and Medway Ports Authority governs the lower reaches extending to Sheerness. These Authorities have jurisdiction over the whole water area up to High Water Mean Spring Tide (H.W.M.S.T.). This tide level should not be confused with Trinity High Water (T.H.W.) which is an arbitrary level related to Dock Areas but now long since disregarded.

The Authorities have extensive powers. On penalty of a fine no work, however small, should proceed without proper application to them. Rather than try to list every aspect, the best advice is to approach them by a pre-arranged visit in order to discuss the project in question. I presented my projects in this way and was always received with the greatest courtesy and given help and guidance. It is important to remember that these Authorities are concerned with every aspect on the river.

The Port and Water Authorities sell Tide Tables to the public. These can be your "bible" as they predict daily tide levels and times. The Authorities also have nautical charts of their areas. The Authorities require applications for licences for new works, new dredging, etc., and they charge for encroachment, that is, where a new frontage extends beyond the existing line.

There are a number of Chart Datums to which tide levels are related; for example, along the Thames and Medway they vary from 3 m to 3.20 m below Newlyn Datum. An elementary sketch or table is recommended to relate these levels in turn to Newlyn and vice versa (see Section on "Soundings" Page 13).

The Area Water Authorities can have jurisdiction over pollution, controlling flood prevention and work within 15 m of the flood prevention line. Large areas of London, Essex and Kent are below high water level. A breach in the flood defences could bring heavy damage claims on to the wrongdoer; be warned and approach the Authorities before undertaking your scheme!

As an example it may be that the reader may intend to drive piles, say at the toe of a river wall, well beyond the 15 m referred to above. In such a case the flood prevention authority (usually the River Authority) would still be concerned with the stability of the wall.

There are maintained flood defences along the banks of the rivers. They consist of a variety of embankments, from clay walls to sophisticated steel piled and concrete frontages. These must not be breached or used as moorings without prior consent from the authority concerned.

In the case of earth embankments particularly, the water authorities will almost certainly only allow bored piles to be sunk for foundations or shore moorings. This type of pile should prevent displacement of material. When breaching such an embankment, the water authority may only allow half the length of the breach to be carried out and only then, most probably, in a properly designed cofferdam.

No work should be carried out on a jetty handling oils and/or spirits before permission is sought from a responsible person attached to the installation. Craft must not be brought alongside without this permission.

"Cold" and "hot work" permits are usually issued daily by the client's Fire Officer, but these can be withdrawn, without notice, and work must cease immediately. Permits normally last for 24 hours at the most.

Many other Regulations are referred to in the text and apply to their particular section.

2. Natural problems of tidal waters

Most tidal waterways have carved their way for many centuries through the terrain searching for the sea, with the result that they are set in their habits and resent man's interference.

Tide flows are rarely parallel to the banks for any distances. They direct themselves from one bank to another, coping in their own fashion with their bends and leaving shoals and channels in the bed. In short lengths tide flows can eddy, giving the effect of a reverse flow.

With such characteristics they endeavour to sweep away any attempts to obstruct their flow by construction workers and frustrate attempts to excavate in their beds by filling the depressions with silt etc. to regain the original contours.

A passage from Philip Howard's "London River" refers to the very old London Bridge and reads:-

"The tide which was fast ebbing, obstructed by the immense piers of the old bridge, poured beneath the arches with a fall of several feet, forming in the river below as many whirlpools as there were arches. Truly tremendous was the roar of the descending water and the bellows of the tremendous gulfs, which swallowed them for a time, then cast them forth, foaming and frothing from their torrid wombs."

Wave action can be increased or decreased in a tideway by the direction of strong winds. Wind opposing the tide flow (wind over tide) will cause a "steep sea", whereas the same wind will flatten the water on the return tide.

It is my experience that wind will increase in a crescendo from low water to high and abate on the ebb. Rain will follow the same pattern. The calmest or best part of a day is invariably during the low water period, if only for a few minutes.

Siltation

This is virtually an unsolved problem in rivers due to the amount of suspended material settling in still water. It seems that however much accumulated silt is taken from an area, or bank of silt, the contours will reform in due course. It would appear that rivers refuse to be altered unless some particular structure is placed nearby.

When considering a proposed structure in the river, one must take into account the possiblity of silting a neighbour's frontage or even the site to be developed.

The surest way to collect silt is to excavate a hole in the river bed. The area will collect several centimetres of silt a day until the original contour is achieved. A sunken barge, with an open hold, will collect some 150 mm per day. Never form a berth lower than the surrounding contours.

Silt is drawn in by water intakes and causes problems in the smaller diameter cooling pipes causing 200 mm diameter pipes to be blocked solid.

Very young mussels are often sucked in and then manage to adhere to the main intake culverts where they grow larger. When chlorine is introduced to the supply to kill off seaweed, the mussels are killed and released. They can continue along the cooling pipes causing 20 mm diameter pipes to be blocked solid.

By using isotopes added to dredged material, it has been established that the material dredged in the upper reaches and dumped at sea near to the river's estuary has returned to its original position within days.

Silt below water line can be cleared in local areas by a diver using a 250 c.f.m. compressor with airline and jet connection. This method is quite effective in "hosing" away material from an area of work. The air jet will lift the silt deposited by the movement of water (and rising air) but care must be taken not to encourage silt to drift beyond the client's own property. Water jets are not so effective underwater because they do not have the lifting effect of the air.

Many schemes have been devised and used to keep a large area of foreshore or "dry" berth free of silt. A reservoir of sufficient size may be built inshore and allowed to fill, on the rising tide by means of a culvert and tide flaps. The water is then released, at low water, through a series of pipes and sluices, to flush the area required to be cleaned. Even where the area to be kept clean is narrow, my experience is that the discharged water merely cuts deep grooves in the silt and mud, leaving the rest undisturbed. This method also has the disadvantage of the reservoir water bringing in suspended silt, which settles out once the intake ceases, so that you finish up with two lots of silt instead of one!

Tide range

The tide range varies considerably in rivers between neap and spring tides. The range is generally higher upriver than downriver, and rises (in the Thames and Medway for example) some 5.500 metres to 8 metres above Chart Datum.

River Authorities issue tide tables which give the predicted time and height of the daily tides. Tides are later each day in a somewhat erratic fashion but take 14 days to go round the 24 hour clock. There are generally two high waters and two low per day.

As a general observation the highest tides are the third after the new and full moon or in the Thames, those tides nearest to 3.20 p.m. G.M.T. at London Bridge. March and October tides are known as the Equinoxes and produce the highest range of heights in the year.

The heights of predicted tides are related to the Chart Datums in the sections of rivers known as Reaches. Chart Datums are included in the Tide Tables and shown on Charts with their relationship to Ordnance Datum.

The slack water on Neaps lasts a little longer than the Springs and is a more suitable time for diving work. Low water springs will expose items at a lower level. A strong sustained wind blowing downriver will drive the ebb tide to a lower level, whereas a sustained "upriver" wind will increase the height of high water.

Oil pollution

This is an offence according to the River Authorities and is subject to heavy fines.

A jetty or quay handling oils or spirits must have adequate fire prevention equipment. There must also be a 150 mm high upstand all round to contain spillage; all drainage from the deck must be collected into a storage tank(s) under the deck. The tank(s) must be fitted with self-operating pumps to discharge the oily water ashore.

Some jetties have a boat standing by permanently, with detergents and spraying equipment aboard, to dispose of smaller spillages on the water. Widespread contamination is likely to be dealt with by attendant craft called in as an emergency operation by the Authority.

The reader is reminded that before introducing a discharge pipe into a river it is necessary to obtain the Authority's approval, but special consideration should also be given to the incorporation of an interceptor for oily water or for preventing toxic waste from escaping.

Care must be taken to prevent diesel storage tanks and plant leaking pollution overside from working craft. All sewage and foul water must be brought ashore for disposal, as is the case with any debris whatsoever.

3. Navigational problems of tidal waters

It is common practice for craft of all sizes to berth with the stem towards the tidal flow, thus using the movement of the water as a brake. To achieve this position, river traffic, which invariably travels in the direction of the flow (to improve speed and save fuel), will require to turn 180° to berth safely.

When choosing a site for a jetty, wharf or any mooring situation, the reader should have regard for the turning circle required for this manoeuvre. It may be necessary for the craft first to proceed beyond its destination to find a reach of river wide enough for the turn round. A berthing area should provide a safe haven for a craft. Thought should therefore be given to the proximity of passing craft, dangerous bends, and the direction of tide flow adjacent to the berth line. There should, of course, be sufficient water for arrival and departure.

The effect of a passing vessel on a moored craft, when the latter is afloat, is to push the stationary craft away with the bow wave, so possibly causing damage to the fendering against which it is lying. Following the bow wave moving craft will then draw off water to replace the bow wave taking the stationary craft with it, perhaps breaking its mooring ropes. This action can then oscillate over several movements and be accompanied by large waves that will, for example, sweep a diver off his feet.

On many occasions, when working in an estuary, I have been on a moored piling craft buffeted in such a way by a passing liner. The crane jib would lurch; the piling hammer swing wildly; the deck heave; moorings snap and men fall about. All this is alarming, particularly at night. One should keep an eye open for all passing craft and bear in mind it can be, perhaps, twenty minutes for the turbulence to reach one's own craft.

One should also make allowances for river bores such as those occurring in the River Severn and Seine.

A ship should be regarded with some special consideration, whatever its size. Rather like women – they are fickle. They are liable to arrive late at their destination by perhaps minutes or days and, when they do arrive require all the attention and comfort demanded by a haughty lady. From a wharf or jetty owner's point of view, these craft are his life line, importing or exporting his goods and he requires knowledgeable construction personnel to be sympathetic to the whims of the visiting vessel.

If any craft is to ground at a berth then care should be taken to ensure that this is safe, either by taking levels (if the berth dries out) or by soundings. Similar precautions should be taken for the approaches if necessary.

It is standard practice that any craft under tow by a tug in the London river requires a licensed lighterman to be aboard the towed craft. His duties include unmooring the craft, then staying aboard and mooring it (with the owner's ropes) at its destination. These men are a law unto themselves and decide how the operations should be carried out and if it is a large pontoon, whether or not more than one man is required to be employed for the work. The crew take no part in all this unless the lighterman calls for assistance.

When arranging a move with a tug company it is prudent to ask them to hire in the lighterman for you, from a lighterage company. At least then the lighterman and tug arrive at the same time and are in accord with each other. If the reader has his own tug, it is imperative that he ensures he is not contravening the practices of towage and personnel. Some construction companies carry their own full-time lighterman but the implications of this practice should be fully realised.

Finally, due deference should be afforded to lightermen as they are subject to a five year apprenticeship and are masters of their calling.

Berthing or building line

When considering a construction projecting into the river, there is a general building or berthing line beyond which jetties and similar structures must not project. All along the tidal rivers is a navigable channel, clearly marked on the Authorities' Charts, and construction must be kept well clear of this. On referring to a chart, the general building line, although somewhat arbitrary, is fairly obvious. Talk to the Authority first to ascertain the limitations.

Underwater construction, such as a pipe or cable, may extend and cross the navigable channel provided there is sufficient cover (usually one metre minimum). Check the depth to which the Authority is liable to dredge; this may extend well beyond the existing bed level. In any event the Authority will normally require the pipe, etc. to be laid below their limiting depth.

Bulbous bows

These are fitted to larger craft at around water line to improve speed and efficiency. The bulbous bows extend further than might at first appear, when the craft is loaded.

As far as jetties, etc., are concerned and when a craft is berthing incorrectly on a "steep" approach line, the bulbous bow can pass under the jetty deck causing damage to supports below the water line. Such damage can be increased by the craft leaving its bow under the jetty and attempting to come alongside; the leverage against piles in an outward direction can be considerable.

A bulbous bow has been known to demolish several piles under a jetty, allowing the deck to collapse. Doubtless a jetty could be designed to avoid such contact, but I can suggest little else, except the utmost vigilance to safeguard against such calamities.

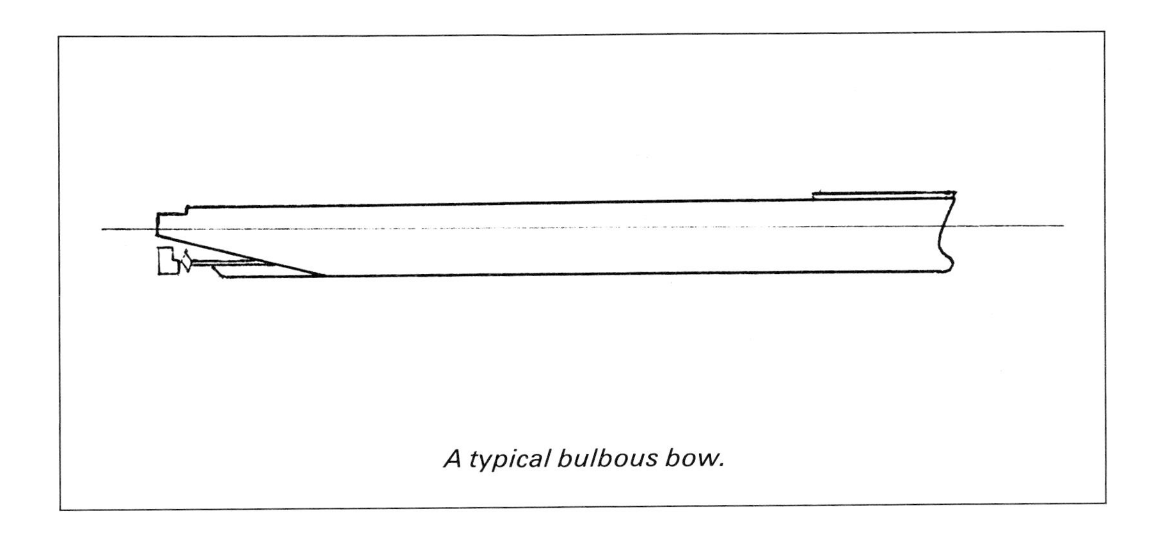

A typical bulbous bow.

Damage to structures

Damage by craft to structures is a frequent occurrence. The unaided "coaster" is more likely to cause damage when berthing at a jetty or wharf than a super-tanker aided by tugs, although of course, when a large vessel takes charge of its escort the effect on a structure is unlimited.

Fendering is usually provided to reduce damage to the main structure and this aspect is dealt with under that heading. (Note: although the craft may have a Pilot aboard to advise the Captain on his handling of the craft, the Captain is answerable for the damage.)

The usual procedure when damage occurs, and provided it can be established a particular craft caused the damage, is that the owner of the damaged structure should contact the Ship's Agent and call a joint site survey as soon as possible, preferably the same day. The owner should also establish a formal claim in writing.

Those attending the survey should be, (i) a representative of the owner of the structure, (ii) his technical surveyor (if the owner's surveyor has sufficient authority, he can act for the owner as well), (iii) a technical surveyor for the ship's Underwriters, and (iv) a contractor specialising in marine repair work, who can give a price on the spot if the damage is minor.

The object of the survey is to agree, entirely without prejudice to liability or other claims, the extent of the damage and cost of repair. The survey may require divers or other technical advice; if this is necessary responsible surveyors will usually agree that additional information is required over and above the visual inspection.

The scope of the survey can be limited to a "friendly half hour" to settle all matters regarding extent of damage and cost, or extend into a major contract requiring Bills of Quantities and competitive tenders. An owner can insist on a Contractor of his choice but the financial aspect needs prior agreement with the Insurers. The survey does not extend into the legal aspect at all. If the Ship's Underwriters deny liability then a legal battle can ensue and a court case may follow.

Whatever the extent of the damage a surveyor should, without fail, treat the matter with the utmost diligence as he may be called as a witness in a Civil Court to justify costs and give reasons for particular actions. The surveyor can be called on to give professional opinions; for example, if assuming that the vessel in question had approached at a reasonable speed and angle, whether or not he considers the jetty was designed to receive such a craft of the particular size invited to berth against it.

However, after all the foregoing, the craft is only liable to pay maximum damage costs of a fixed rate times the Registered Tonnage. The remaining costs are borne by the owner of the structure so it is imperative that he insures for a total loss brought about by craft, in addition to his usual cover for fire, etc.

Certified engine failure and "Acts of God" are only two of the many reasons thought up by Underwriters in an attempt to avoid liability.

My advice is that when attending a survey on behalf of the structure's owner, it is advisable to address oneself only to the ship's underwriter's surveyor. Never get embroiled with the ship's captain (or anyone else) and particularly not with foreigners. The underwriter's surveyor will need all the information he can get from the captain in order to report to his Principals on how the craft was behaving immediately prior to impact.

Propeller damage to berths

This section refers to berths where craft take the ground at low water.

There appears to be no material from which to make a berth (except concrete) which will withstand the effect of a ship's rotating propeller. Ships which take the ground are fairly small and berth without help from tugs, so that the propeller is the only method of moving forward, or astern, or braking.

All ships' masters are impatient and all wharf or jetty owners want to berth larger and deeper ships. The result is propeller damage due to lack of water under the vessel. Often a master will order "hard astern" to avoid heavy contact forward. This action can easily scour out 50 to 100 cubic metres of material from the berth, leaving a crater, and throw the substance forward under the craft. The craft will then ground on the heap it has thrown up, causing bottom damage and a claim from the ship's owner for a foul berth. Owing to the crater the engine area can also be left unsupported.

The reader should refer to the section on "Berths" (see Page 51) for advice on regular surveys of berths and to "Damage to Structures" (see Page 11) for claim procedures.

Obviously a claim for a foul berth will be by the ship against the wharf or jetty owner. The surveyors may require to dry dock the ship to make sure the shape of the damaged bottom fits the survey made of the alleged foul berth.

A ship's owner may unjustifiably claim a foul berth and resulting damage in the hope that bottom damage, caused by the master striking a dock sill elsewhere, will be paid for by an unwary wharf owner. A wharf or jetty owner should have no hesitation in claiming berth damage against a ship which scours his berth.

A wharf or jetty owner should ensure that any tide gauge he provides on the side of the jetty or wharf for the benefit of the ship's master is a correct indication of the amount of water over the berth. Crucial questions can enter into the case such as "Was the ship invited to berth when there was insufficient water?" or "Did the ship berth without permission?".

From what has been written here the reader may consider that berths and bottom damages can be a nightmare and certainly I always disliked being involved in such cases. The intricacies of litigation are far more involved than structural damage.

Concrete berths may seem to be the answer to all these problems, but unfortunately much large debris can be washed on to a berth, and any hard object will rather cause instant damage to a ship's bottom than be pushed into softer material. Concrete berths should generally be considered impracticable except in very special circumstances.

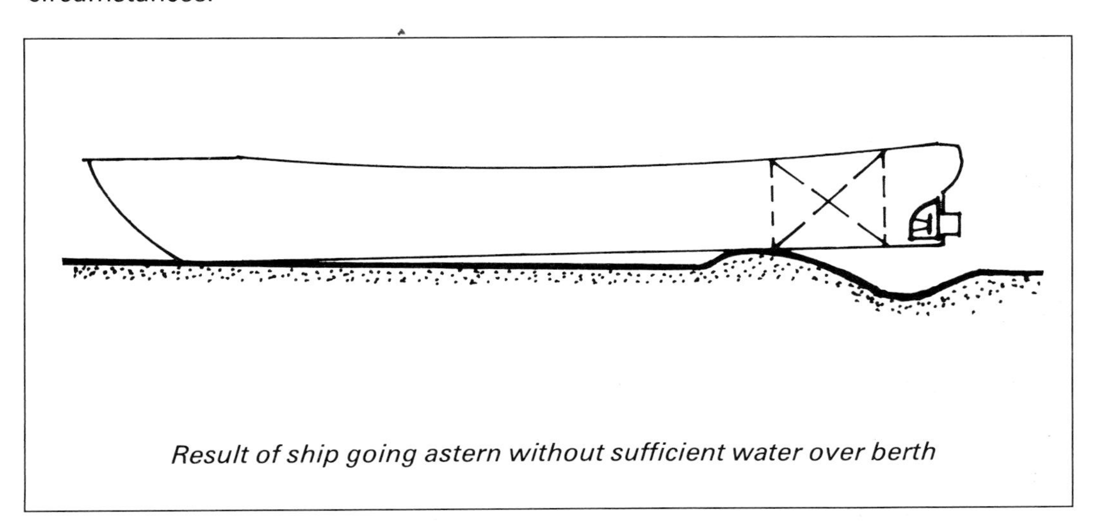

Result of ship going astern without sufficient water over berth

Soundings

Generally, if soundings relating to an area are required most river authorities have up-to-date charts to which the reader can refer. If a particular area needs a more detailed sounding, and one is not able to carry out the survey oneself, it is recommended that the river authority is engaged. They are experts and produce authenticated results at a moderate charge. There are also freelance surveyors who carry out the work.

All the above refers to echo soundings which are usually based on the local Chart Datum (see section on "Regulations" Page 4). To ascertain the actual depth of water one should refer to the river authorities' Tide Tables for the particular area. In March and September there may be the occasional "minus" low water, but rarely does low water get as low as Chart Datum.

Soundings can be related to a particular berth level, so that the owner can easily relate a predicted tide level to the depth of water on his berth and a tide gauge can then be fitted accordingly. If asked to fix a tide gauge, always keep the numbers "sitting" on the line; it doesn't do much harm to keep the board up 300 mm or so high just to fool the impatient skippers referred to earlier.

Soundings can also be carried out by using a pole marked off in cms and metres (and multiples thereof). Use aluminium if possible about 37 mm diameter and in 2 metre screwed sections. Over 6 metres overall such poles are not practical to handle. Always use a dinghy rowed by a waterman; never use a motorboat. One needs a distance line from a fixed point to establish the position of each sounding and a motorboat is too fast, uncontrollable and dangerous to the crew, to have a fixed line out. Someone is needed to do the booking and another to control the line. A third person is required to "dip" and read the pole and call out the depth at each station or distance. Thus a crew of four is needed for the work.

The alternative to a sounding pole is a proper sounding chain or Navy type lead line. I had the experience of using a Sutcliffe Wheel when sounding on the Manchester Ship Canal. This consists of a wooden spoked wheel about 750 mm diameter with several turns of sounding wire running over it, in grooves, down to a lead weight. The wheel is set up over the stern of a small boat. From the hub is mounted a spiral which also has a wire which in turn is also connected to the lead weight. The spiral is so constructed that it pays out an hypotenuse to the vertical wire coming off the wheel. This arrangement holds the lead plumb and prevents trailing due to the movement of the boat through the water. By "feeling" the lead along the bottom (by rotating the wheel slightly in each direction) and knowing that it is being held vertically, it is possible to carry out very accurate soundings. The soundings are indicated by a fixed pointer against the depths marked on the alternating wheel (clockwise and anticlockwise) handled by the operator.

I consider that this very old fashioned accurate method should be used more widely.

One advantage the sounding pole has over all other methods is that not only can the top of the soft mud be felt (which is the limitation of echo sounding, sounding chain and wheel methods), but the pole can also be depressed to find the firm ground below. This can be very important when sounding a berth where craft are to ground, as it draws attention to any hazardous irregularities which could damage the ship's bottom.

In all cases one needs a tide gauge or board within view to refer to frequently in order to establish the varying tide levels during the survey. The board's "zero" should be set to a known datum to which the soundings can be related.

If only a minimum depth is required over an area, and to make sure there are no isolated high spots or obstructions which could be missed by a grid system, a sweep is an ideal apparatus. This consists of a raft (or similar) some 5 to 7 metres wide, with sufficient buoyancy to carry the weight of a bullhead rail suspended from the raft, on a wire each side, at the required level. The rail should be kept in position fore and aft by two pairs of "hypotenuse" wires back to the raft. The raft is then towed by a small motor boat and towing bridle. Any obstruction can be felt by the reaction of the boat or the sudden erratic movement of the raft.

Unless the vertical distance between raft and rail can be varied, the operation should be carried out at low water and over a period when the water level does not vary. Do not forget to relate the tide level at the time of the survey with the tides concerned with the forthcoming ship's movements.

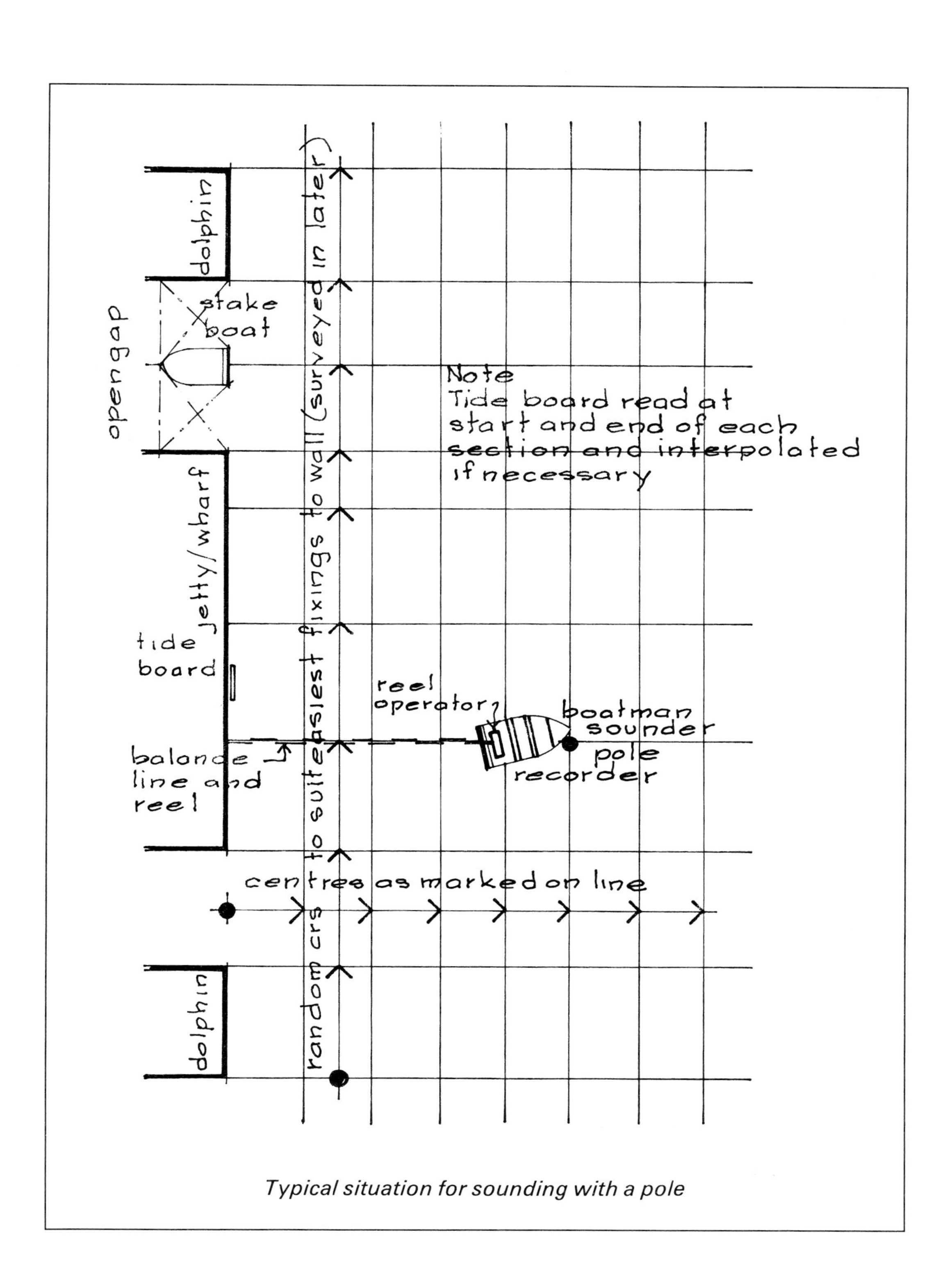

Typical situation for sounding with a pole

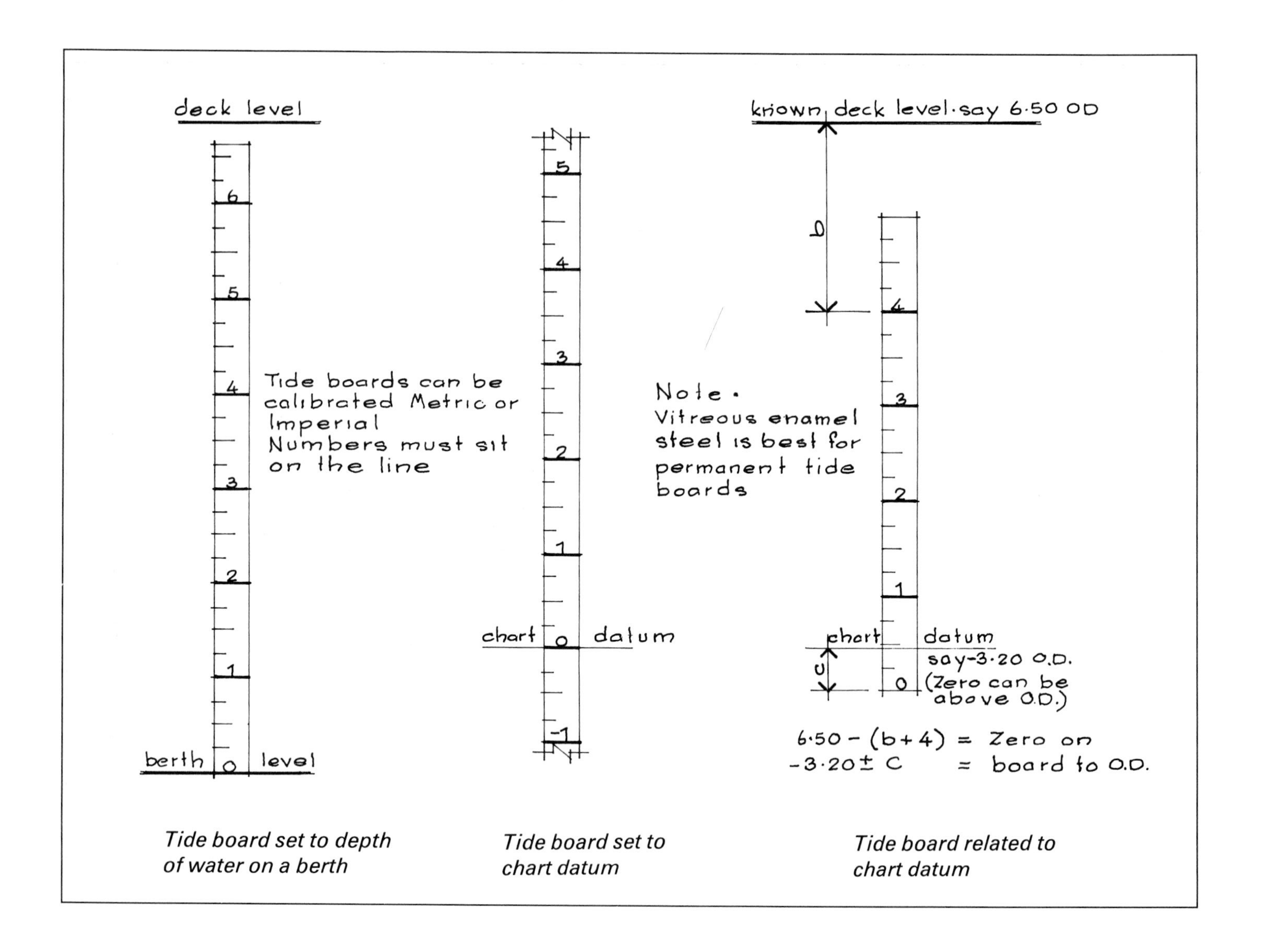

Tide board set to depth of water on a berth

Tide board set to chart datum

Tide board related to chart datum

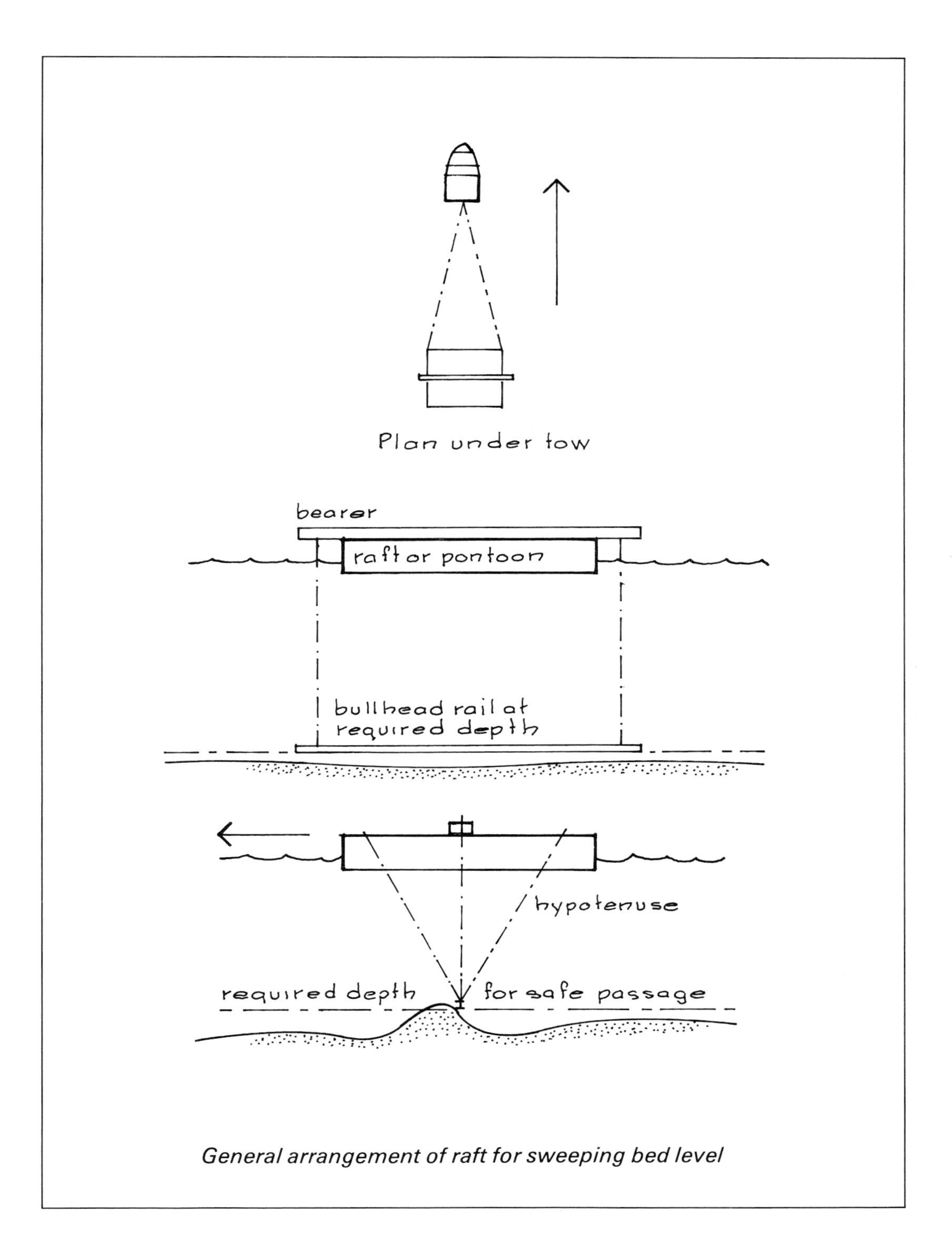

General arrangement of raft for sweeping bed level

Tidework

Usually this is resorted to when only low water work is available to a working party. It may incur overtime rates if the working period falls outside the normal working hours. Double tidework means working both (low water) tides in each day. More often than not the work pattern is, nominally, five hours worked continuously followed by seven hours off, but with due allowance for the second tide being later than the previous one. Tidework rates apply to such work.

My personal experience is that one week of such hours is tiring to the workforce and a weekend break is welcome. To work two weeks without a break is about the limit of acceptable endurance before output flags. After such a period, one sees the sun on the horizon and wonders if it is setting or rising (sometimes it can even be mistaken for the moon) and one's stomach questions the irregular hours of meals.

The main purpose of this section is to draw attention to the fact that, due to the short working period, output should be at a maximum. It is, therefore, better to spend the last hour of a working tide preparing for the next, so that the subsequent working period starts without delay.

During single or double tide work sessions the tendency of the workforce (and divers) is to arrive less than 2½ hours before low water and to slacken further in this timekeeping as the days go by; they have a fixed notion that low water is starting time. Unfortunately once the tide turns all can be lost. It is imperative that supervision on sites ensures the best of timekeeping, and that plant, material, lighting, etc., are in efficient order and ready when required.

4. Plant and materials in river construction.

It is fortunate that the main ingredients of marine construction are but three; timber, concrete and steel.

Up to about 1920 timber was used extensively for jetty, wharf and cofferdam construction. British Columbian pine was readily available in very large section and long lengths and was considered to be cheap, even in those days, at £2.10 per cubic metre. Timber structures were easily repaired in this material even to the extent that if a pile broke off below bed level another could be driven alongside and the damaged walings and bracings stitched together somehow.

However, all softwoods in contact with a river bed are subject to attack from the toredo beetle. These creatures, once limited to warmer waters, have been attracted to the U.K. over the last 50 years or so by the warm water discharged from cooling water systems. Without detection from outside these beetles bore continuous tunnels, the diameter of a lead pencil, along the grain of softwood until suddenly, perhaps, a pile will collapse. The tunnels are left lined with a grey substance.

Another area for attack is around the high water area, known as "the wind and water line". Here the timber is subject to total decay in 10 to 15 years due to the continuous wetting by the tides and drying out by the wind and sun. This area is a haven for timber lice of enormous proportions, which live off the rotting timber. My hair would curl as I descended a vertical ladder over the side of an infected wharf, and came face to face (within inches) with these vile things. As a contractor's man there was the consolation that they were providing future work!

Elm is subject to fairly rapid decay at the wind and water line, whereas greenheart timber is the best to withstand all the above described attacks.

To achieve the best results, careful planning must be given to the conveyance of concrete over water or to its mixing on a craft adjacent to the shutters offshore. Pumping will no doubt require staging to carry the pipe line and this should be wide enough to allow access to the pipe. Trying to pump concrete over 160 metres through a 150 mm diameter pipe, on a hot summer's day, was a disaster for me. The concrete jammed in the pipe line and set hard, writing-off the pipe, concrete and the day's labour and plant. Pumping downhill is not satisfactory.

Concrete can be mixed ashore and conveyed in skips by barge or pontoon. Retardents may be required to extend the placing time. Also, sufficient water to float the barge or pontoon at the transfer positions is essential, so watch the tide levels during the essential placing period.

Concrete can also be mixed on a craft once the aggregates, cement and water are there. If fresh water supplies are difficult to obtain from the shore, there are motorised water barges available, on larger rivers, which sell it by the tonne. All the above operations in concrete will require some form of cranage.

Steel does not present much of a problem itself, apart from corrosion which is dealt with under "Painting". The reader should guard against mixing steel and timber in permanent free standing structures. There is always some movement in such a structure, however slight, in a tideway. Holes in timber for bolt fixing purposes are always drilled at least 3 mm oversize though in steelwork usually to a much tighter tolerance. This allows movement and wear between the bolt and the timber, which gradually causes the hole to enlarge and the whole fixing to become sloppy. As a consequence the structure increases its movement and a sawing effect develops between the bolt and steel. I was once engaged on removing steel channel walings and bracings from a timber piled jetty where all the 32 mm diameter bolts had been partially or completely "sawn" through. Some steel members had fallen off completely; others had enlarged bolt holes. Timber to timber is satisfactory provided that bolt tightening is maintained, say once a year.

The reader should enquire of the nature of materials handled by the owner of a local existing structure to consider if they are liable to affect his own construction materials.

Pneumatic tools under water

The average pneumatic tool used underwater has its efficiency greatly reduced by the pressure of water the exhaust has to overcome. If possible, an exhaust line should be introduced to discharge above the water level.

Brickwork in tidal conditions

This is to be avoided if at all possible. Bricklaying and pointing, if unavoidable, are best carried out on an ebb tide to give the mortar some time to go off before the flood tide. The smallest of waves slapping on the new work will wash out any unset mortar. The working hours are therefore limited and should be programmed accordingly; if necessary tidework hours should be introduced.

Painting

It is my view that extravagant claims for their products have been made by paint manufacturers. In my opinion no modern paints last long enough to warrant the cost of purchase and application – particularly maintenance painting in tidal conditions. It is far cheaper in the long run to let corrosion occur. Anyway a steel-piled wharf front can only be painted on one side after construction. Up to about 1950 it was customary to keep steel in stock until the mill scale loosened to be wire brushed off. Then genuine red lead paint was applied, followed by two coats of coal tar from the local Gas Works. It is my experience that this application lasted longer than the modern ones.

Shot-blasting has taken the place of waiting for the mill scale to loosen and it is very effective in cleaning steel, provided the weather conditions allow and a primer is applied immediately afterwards.

Incised and pressure-creosoted softwoods have a longer life in water, but the reader has to reckon on decay setting in at the "wind and water" line after 10 to 15 years. If timber fender piles are vulnerable to damage, he has to consider if creosoting is worthwhile, bearing in mind the possible natural life span of the fenders.

There is no value in treating hardwoods with preservatives as they will not penetrate the surface area sufficiently.

Sugar

Sugar is absolutely fatal when in contact with wet or newly placed concrete. A "sugary" atmosphere, say at a sugar refinery or jetty, is sometimes sufficient to cause failure in the setting of concrete. Even when tarpaulin covers have been placed over new concrete, I have experienced failure in setting because sugar from overhead loading has dropped on to the covers. Aggregates can also be contaminated, with the result that the later concrete is ruined.

When the reconstruction of a Tate and Lyle's jetty crane-track was carried out, the adjacent areas and meeting faces were washed down and steam cleaned before the placing of concrete. This met with success. As an added precaution no jetty activity was carried out during concreting.

Types of timber

For the purpose of river construction work timber can be classified into softwoods and hardwoods.

(a) **Softwoods**:
British Columbian Pine (B.C.P.) Douglas Fir and Oregon Pine are the same timber. It is useful for general timber work down to 75 mm × 225 mm in section, its weight being approximately 620 kg/m^3.
Pitch Pine is supposedly a more durable and stronger wood that the other pines and fir referred to above. It is my opinion that in recent years supplies do not meet this better specification and this particular timber is not worth the extra cost. Its weight is approximately 930 kg/m^3.

(b) **Hardwoods**:
Greenheart – either sawn square or hewn in the forest to roughly square and tapered from butt to tip. A very durable and straight grained timber with hardly any knots and resistant to toredo beetle. Its weight is over 1300 kg/m^3, so will not float. Very long delivery periods can be expected.
Opepe – a very good alternative to greenheart for shorter lengths.
Jarrah – an Australian redwood with similar qualities to greenheart. Only short lengths are available, its weight approximately 1225 kg/m^3.
Oak – is a good durable timber suitable for rubbing pieces as fendering. Its weight is 950–1000 kg/m^3.

Elm – similar uses to oak but spongier and not so durable. Available in short lengths and becoming less available due to Dutch Elm disease. Its weight is approximately 900 kg/m³.
Ekki – suitable for decking timber; hard wearing and durable. Its weight is approximately 1200 kg/m³.

Note: Hardwoods will wear away softwoods if allowed to chafe against them.

Fire damage to timber structures

This occurs only too frequently, particularly when any acetylene burning gear or welding gear is used. The draught around a jetty fans any smouldering timber which may not ignite fully until several hours afterwards. In the case of a fire to a holiday resort pier theatre, arc welding equipment was believed to have caught the building alight and the fire extended from end to end of the building in three minutes. Sufficient standby fire-fighting equipment is essential but regrettably this priority is often ignored.

Fires rarely go much below deck level because the next high water will extinguish anything burning in the tidal range. Tugs may come to the rescue with fire-fighting hoses, but they can make a charge for their services.

Do not have men so placed that they cannot escape from a fire. Have a safety boat on standby and a ladder suitably placed to get to it.

Repairs to concrete structures

There was a period when reinforced concrete jetties were the mode and they were designed in a similar manner to structures ashore. Those constructed over the water nearly all suffered from insufficient cover on the reinforcement, particularly to the undersides of the slabs, where condensation occurs, with the result that the reinforcement rusts and bursts off the concrete. By this time the reinforcement is dangerously wasted away.

Trading companies offer a service to remove the remaining spalling concrete and jet on a replacement in "dry" sand and cement mix. This method has often been applied by "trade names" at considerable expense, to "repair" the spalled concrete. It is my experience that after a number of years the sand/cement mixture detaches itself from the original face and if it does not fall off it leaves a hollow surface. In the meantime the corrosion of the steel continues.

My conclusions are:-

(i) the method of repair described above only adds weight to the already weakened structure. The designed profile is often exceeded
(ii) the only proven method of repair is to fully expose the steel, cut back behind it, and clean it off with acid; then use a bonding agent to the newly exposed surface and apply shuttered concrete. Additional reinforcement should be fixed to make good the loss due to corrosion
(iii) no new reinforced structure in the river should have less than 75 mm of cover over the links

(iv) the most elementary source of spalling is the binding wire offcuts lying on the soffit shutter and then allowed to be concreted in. If these are left in the exposed soffit surface they start to corrode immediately. With a cover of 37 mm on the main bars the links are soon attacked, and this is followed by an assault on the main bars

(v) use eyeletted wire ties or stainless steel binding wire, and above all ensure the soffit shutter is clean before any concrete is placed.

5. Site works

If the reader has a good foreman, used to tidal conditions, then his troubles will be reduced to a minimum. The foreman is the Sergeant-Major of the work force; is capable of carrying out any task; is well experienced and knowledgeable; can handle and get the best out of men; and, not infrequently, he will be quite well educated. With all these talents, he will still need a back up service from the reader.

Beware of the foreman who claims to "know it all" on water, having gained his experience in the Middle-East, where often there is no tideway. He will be lost in tidal conditions.

The physical endurance required in and over water is far greater than ashore. It is invariably 10°F colder, and meal times are not always regular. When arriving on site on a bitterly cold day, well wrapped up for a brief visit, remember that the men have probably been there since early morning and worked two hours past their meal break in order to complete a task. Living conditions afloat are pretty elementary.

The supervisor should be welcomed as a person who has come to the site to give help and support, not (unless necessary) as someone always expecting more to be done, yet lacking in his own endeavours and duties.

The foreman has to think of every detail at close quarters to the work. Consequently, at the best he can only see a few days ahead. The supervisor must ensure that he gets his every need – out there afloat possibly with no telephone – by not only thinking of the day in hand but next week and several weeks after that. Work afloat, in a tideway, is slow enough and at times frustrating, so make sure his plant and material arrive in good time to avoid further delays. Talk things over with him during a tea-break, not while he's trying to pitch a pile before the tide turns.

All marine gear is heavy, even down to a shackle. As a consequence tackle and wires are left where they were last used and are hazards to the crew in a confined working space. Keep the place tidy by words of encouragement.

It is a golden rule in tideway construction to grasp every opportunity to carry out low-water work as a priority, leaving work above more as a "fill-in". Do not make a show by completing all the "top" work in good time only to find that the outstanding low-water work is restricted to an hour or two per day, leaving idle time for the remainder. Watch the tide tables and make the very most of the Spring tide low waters.

The pontoon is lifted and lowered by the tide. Use it, pay for the welders to reach all levels of work throughout the day. Scaffolding is a liability. Employ all the welders you can on the low water work and avoid pressure from them for extra reward should you be left with only this level to work at.

As a young man I was given several pieces of advice:-

(i) Never accept without question someone else's measurement as correct.
(ii) A good surveyor devises elementary checks on his own work.
(iii) If faced with what appears to be an insurmountable adversity, ignore it and proceed to the final stage. Then turn about and demolish the adversity from its rear.
(iv) When required to produce a multiple of an object, make one complete to the last detail, in case of snags, before completing the remainder.
(v) If your boss approaches you, hop off your stool and offer it to him!

My own embarrassing observation is that more time and money is lost in construction work by the mistakes of middle management (of which I was one). Forgetfulness, lack of thought, mistakes in setting out and general *laissez-faire* is far too common at this level, yet these people are the first to pour scorn on a foreman who makes a mistake. Foremen have a lot of common sense – that's why they get the job.

Health and safety is a part of any construction work, but when afloat you should always add a safety boat and man to the list.

When considering the timing of a site visit or making an appointment to view a marine structure always consult a tide table first so that you see it at low water or before, to allow time to see all that is required. It is obvious that when the tide is high you won't see much of the lower areas.

I found that a good deal of time and money was lost, when starting a job, because the foreman and I had not remembered to order some particular tool or material. This error incurred extra transport, delay, or even missing the towing away time.

To reduce this cost I wrote a list of every tool and piece of plant I could think of (in alphabetical order) and another for materials. Both the foreman and I could refer to these without having to think up a new list each time. For many years I had twelve sites running at the same time, with as many foremen, each with his list in his pocket. It worked wonders!

Design

This is a complex subject, made more complex by various assumptions which have to be taken before basic principles are applied. Some examples are:

(i) what is a reasonable "closing" speed of a craft on berthing? – 150 mm per second is a common assumption.
(ii) the angle of approach? – 30° is often used.
(iii) does the craft absorb 50% of the impact, and the structure 50%?
(iv) can the period of contact be extended by flexibility in the structure and/or fendering? – this would allow an increase in the closing speed and/or angle, by reducing the absorption value of the design of the structure. (A choice of fendering is described in the section "Fendering" – see Page 38).

In the case of a mooring dolphin intended to take a pull from one direction only, it can be subjected to a pull, at anything up to 180° from the original direction, as, for example, when a craft overshoots and uses the dolphin as a brake by ropes to its bollard.

The capacity and anchorage of a bollard should well exceed that of the breaking strain of a mooring rope or wire. Wind load against a craft's side, which could be transferred to the structure, should also be considered. The introduction of sophisticated fendering systems can reduce the need for a more stable structure.

Therefore, in starting with the size, displacement and depth of water required for a craft to berth, the reader is then faced with the questions mentioned above. To these he must add common sense, his experience and "feel" for the work and a good deal of hope that, if a calamity should occur, the offending craft will have approached the structure at an unreasonable speed and angle and that his design will have been scrutinised by an engineer who makes different basic assumptions to the reader.

To save a good deal of time, the reader should approach the Authorities informally with an outline plan to see what problems there may be. They may require navigation lights to be fitted, etc., and neighbours will have to be approached eventually.

In siting a structure endeavour to keep the berthing line parallel with the general tidal flow. If the Authority is unable to give the tidal flow directions for the particular locality then the reader should take flow observations at the appropriate berthing times, by means of small buoys and pennants to catch the lower flow direction rather than those on the surface.

There is another important consideration that structures must not extend into the river beyond the "building line" set by the Authority. As a "yard-stick", the outer face of the craft, berthed at a structure, must be well clear of the Navigational Channel or Fairway, which will be clearly marked on the River Charts. (Also refer to the section on Ladders and Handgrips Page 46).

In designing a jetty it is usual to use the displacement tonnage of the craft in the calculations. This is obtained by the formula:

$$\frac{L \times B \times D \times \text{Coefficient of Fineness}}{35}$$

when the Coefficient of Fineness is usually taken as between 0.5 and 0.8. Other tonnages are:

Gross tonnage or Registered tonnage:– Cubic feet of all enclosed space divided by 100

Nett tonnage:– Gross tonnage less deductions of crew space, machinery, etc.

Deadweight tonnage:– The weight which brings down a ship from her light to local draught.

Thames measurement:– From the length (measured from the fore side of stern to the after side of the sternpost on deck), deduct the breadth, multiply this result by the breadth and that product by the half-breadth, and divide by 94 – and there you have it!

form can be seen almost daily in the Sea Dredged Aggregate trade, where craft go to the open sea to retrieve ballast for the concrete producers upriver.

Suction dredgers are capable of carrying large quantities of materials. They are designed to have bottom-opening doors for discharging or, in the case of sea dredged aggregate boats, they are self-discharging by means of scoops and conveyors on board unloading into conveyors ashore. This type of installation is dealt with under "Jetties" (see Page 44).

Before dredging commences, the following procedures are recommended.

(i) Make sure to select a specialist with the right craft and knowledge.
(ii) Make sure that the reduced level of the bed will not jeopardize the stability of nearby structures on the river wall.
(iii) Be responsible for any silt from the operation settling on a neighbour's river frontage.
(iv) Make an accurate survey of the area immediately before work starts and immediately afterwards (or as the work proceeds), and agree with the client or contractor.
(v) Agree a method of measurement. It is customary to dredge below the required level "to make sure" the level is achieved. Most sub-contractors will require to be paid for this dredging. Often the quantity of dredged material is measured in a calibrated hopper – the measurement depending on the draught of the hopper. Remember that the material can be measured "in situ" by "before and after" surveys but the material in hopper will have bulked by about 25% so agree a method of measurement beforehand.
(vi) Adhere to the following procedure which is mandatory. If the dredging is carried out to achieve a new level an application must be made to the river authority and a licence obtained. If the work is only to obtain a previously licenced level, then the reader should refer to it as "removing accumulated material" in order to prevent raising a hare with the authority regarding a licence. In either case the authority will require to know the amount of material removed for the records.

Hoppers are an essential part of dredging equipment. Thames lighters, or any barge, can be used when dealing with a small quantity of material, but it is best to pump away the water brought up and deposited in the hold; otherwise the reader will have an unstable barge and the pleasure of paying for towing away a quantity of dirty water.

Hoppers are barge shaped, with main buoyancy tanks on each side and a centre hold. They can be self-propelled. A longitudinal strong back acts as a stiffener for the craft and is used to suspend chains to doors forming the false bottom of the hold. A power pack operates the chains to close the doors before filling the hold and to open them at the dumping ground.

This system can be used, for example, for transporting rock to fill a hole underwater. Although the hopper will rise out of the water as the rock is discharged, it is necessary to ensure there is sufficient water over previously placed material to allow the doors to open and the craft to clear its contents. This operation carried out on a falling tide is fraught with danger.

A more sophisticated self-discharging hopper is the "split hopper" type. Usually self-propelled, it discharges itself by opening up the hold like a grab, thereby avoiding any necessity for the extra draught referred to above.

Provided the material brought up in the dredger is soft and free of debris, there are installations ashore for pumping out the arisings from hoppers and barges into lagoons.

The total cost of dredging, together with the towage of attendant craft and disposal, can be many times more expensive than common machine excavation ashore.

Driving sheet piling in a circle

Steel sheet piling will always want to "creep" at the head in the direction of the driving. With straight line driving, an experienced piling foreman will know how to obviate this and keep them plumb.

However, when driving a circle, whether on land or in the water, the control is not so easy. The operator should partly drive the first pile sufficiently to support further undriven piles as they are pitched – but lean the first pile away from the direction of driving. Complete the circle and, if his luck is in, the last one will interlock with the first and with all the piles leaning away from the direction of driving. During driving, which should be taken in several stages of penetration around the circle, the piles should become plumb.

If the circle (or any "closed" pattern) is in tidal water, one must leave one or two piles raised to allow the water to flow in and out until the necessary struts (and ties) are fitted. Closure should be made at slack water. Providing "gates", for supporting against the tide flow, and aligning a circle of piles, can be an expensive operation, involving temporary piles and substantial walings.

The piling manufacturer will give a good deal of assistance in choosing the right kind of pile, the method of driving, and the diameter to be achieved. Flat-web piling is sometimes used for small diameter circles, but owing to its flimsy nature and difficulty in handling, one should avoid this if possible.

Driving timber piles

Methods of driving piles are improving every day so I will not risk out-dating myself by recommending any particular procedure. However, when driving timber piles the head must be shaped properly, usually by an adze, to receive a special driving ring of the right diameter. Any compromise will cause damage and splitting of the pile head.

Always use a properly fitted shoe (either "diamond" for vertical piles or "sheeters" for closed piling). These rules apply to both soft and hardwoods.

Timber piles do not like "Thames" ballast. At Coryton and Wandsworth the ballast is at, or near, the surface of the river bed and foreshore. As a timber pile usually refuses to penetrate the ballast more than 600 mm, the stability of the pile must be in question; what is disconcerting is when the pile floats out after driving. If timber piles are punished to achieve a penetration, the shoe will be driven up into the timber. A 25 mm set, using a 2 to 3 tonne hammer falling 1 metre, is about the limit before collapse.

At Coryton I once drove five 21 metre long timber piles at low water as a "bent", 16 metres wide, and immediately fixed walings and bracings to form one unit. Owing to lack of penetration into the ballast, on the rising tide the whole set "welled" up, fell flat onto the water and drifted away. Fortunately the piling barge had been pulled clear, having completed its task for the day.

Extending piles in place

The exercise of extending driven piles is slow and expensive enough on land, but it is far worse afloat in tidal waters.

Firstly one has to convey the necessary plant and material over the water to the piling craft. Secondly, if the piling craft is used as a working platform, the craft cannot proceed with piling at the same time. Thirdly, and probably the worst aspect, is the fact that, with a tidal range the staging provided by the craft is rising and falling, which can limit the working time on welding and formwork etc. considerably. Apart from the material, it costs almost as much to extend a pile 300 mm as a much longer length.

The conclusion to be drawn from this is that it is cheaper to cut off a reasonable surplus length than add to a "short" pile.

Boreholes may give a very good guide to the length of pile required, but if circumstances permit it is better to drive test piles to the required "set" before ordering a quantity "on spec". Rivers have a nasty habit of proving borehole information unreliable.

Pile extraction

This is said to be the biggest gamble in river work. Before attempting extraction the reader should assume the pile has previously been driven to "refusal", or near to it. Never start with lightweight gear in the hope the pile will come out easily.

The river authorities have craft and equipment, often called "Yantlets", which are specially designed for extracting and general lifting; these are the best type to use. Each craft is capable of about 140 tonnes pull, and are used for the purpose, on a rising tide, in order to add additional lift through the floatation of the craft. Any other craft is really an adaptation and no more than the best one can muster at the time.

If faced with using an alternative to hiring the authorities' craft and crew, that is adapting a craft, one golden rule is that the anchorage point, usually a winch, must be able to release the load by means of a clutch or brake. Never insert a pawl, as once the load, coming on to the winch, exceeds the capacity of the winch, the load cannot be released. If caught with this situation, on a rising tide or heavy swell, the result is a broken winch, the winch torn out of the deck or, if the strop round the pile holds, the craft will be held under the rising water. It is a brave person who cuts the strop with burning gear in these circumstances.

Timber piles will only take "so much" from a wire strop before decapitation. Get the strop as low as possible and make the pull vertical. Concrete piles usually shed their concrete cover, exposing the main reinforcing bars, and crush. These piles are difficult to extract. A wire strop on steel piles will slip; always insert timber packing between the two.

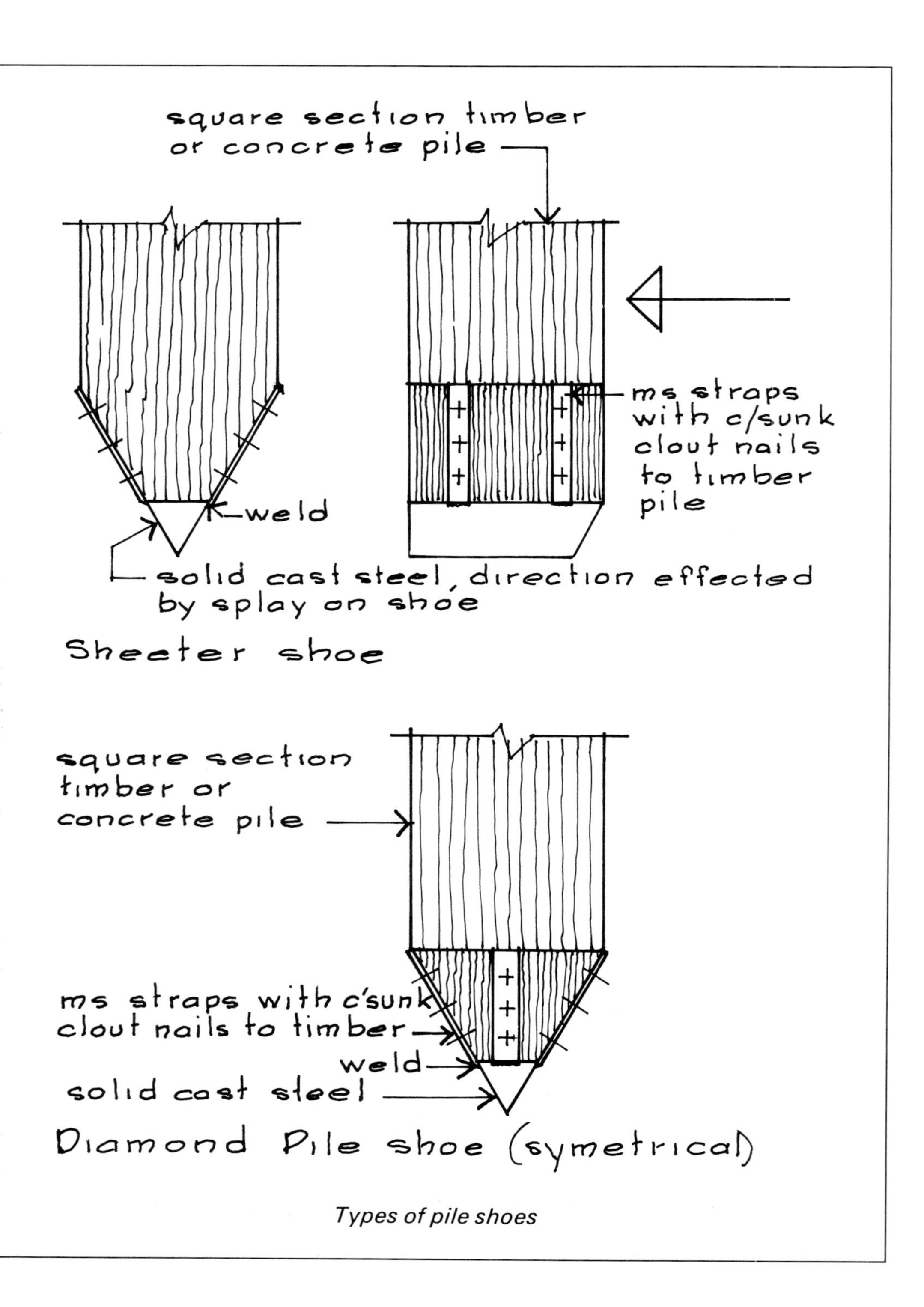

Types of pile shoes

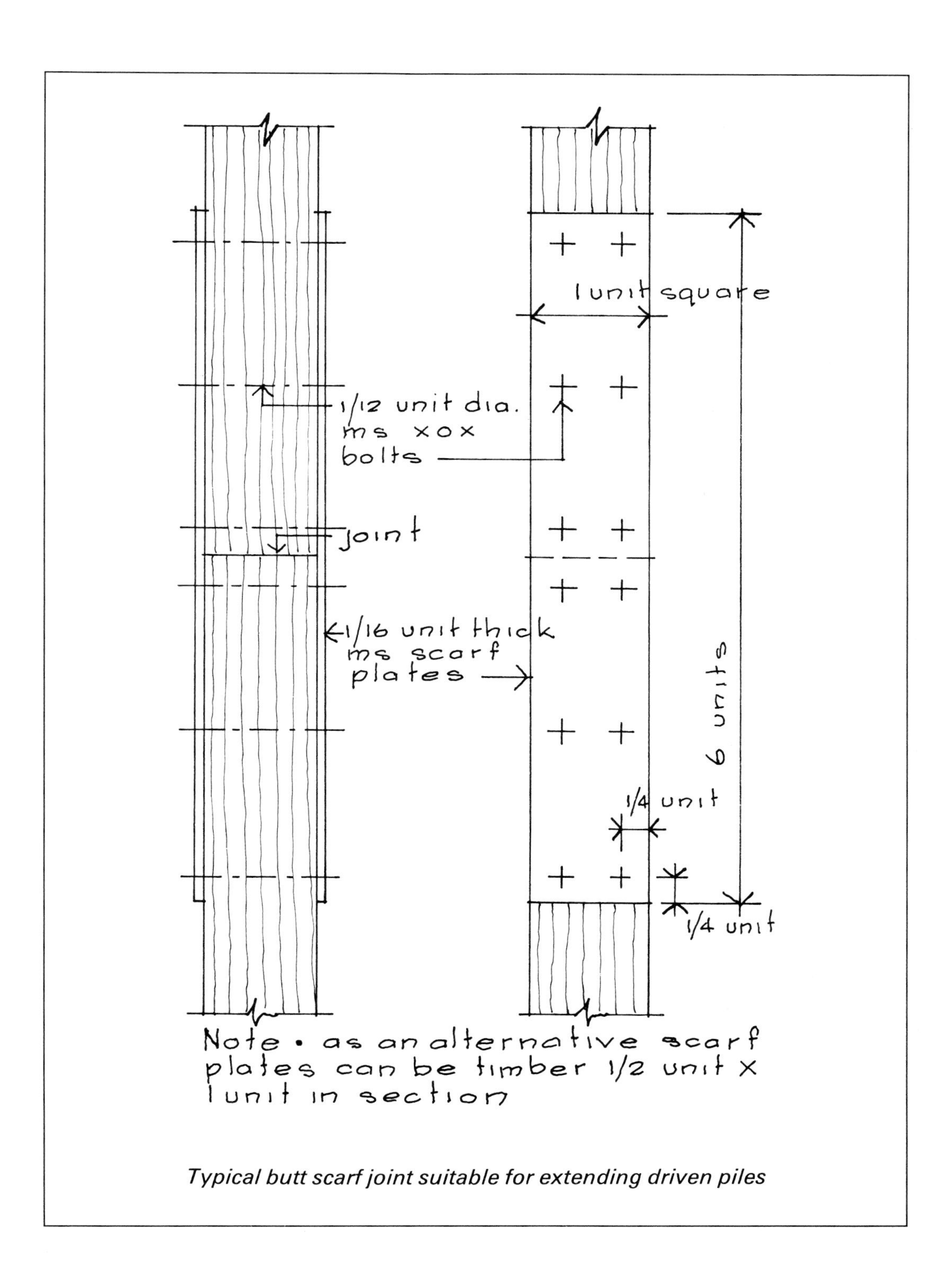

Typical butt scarf joint suitable for extending driven piles

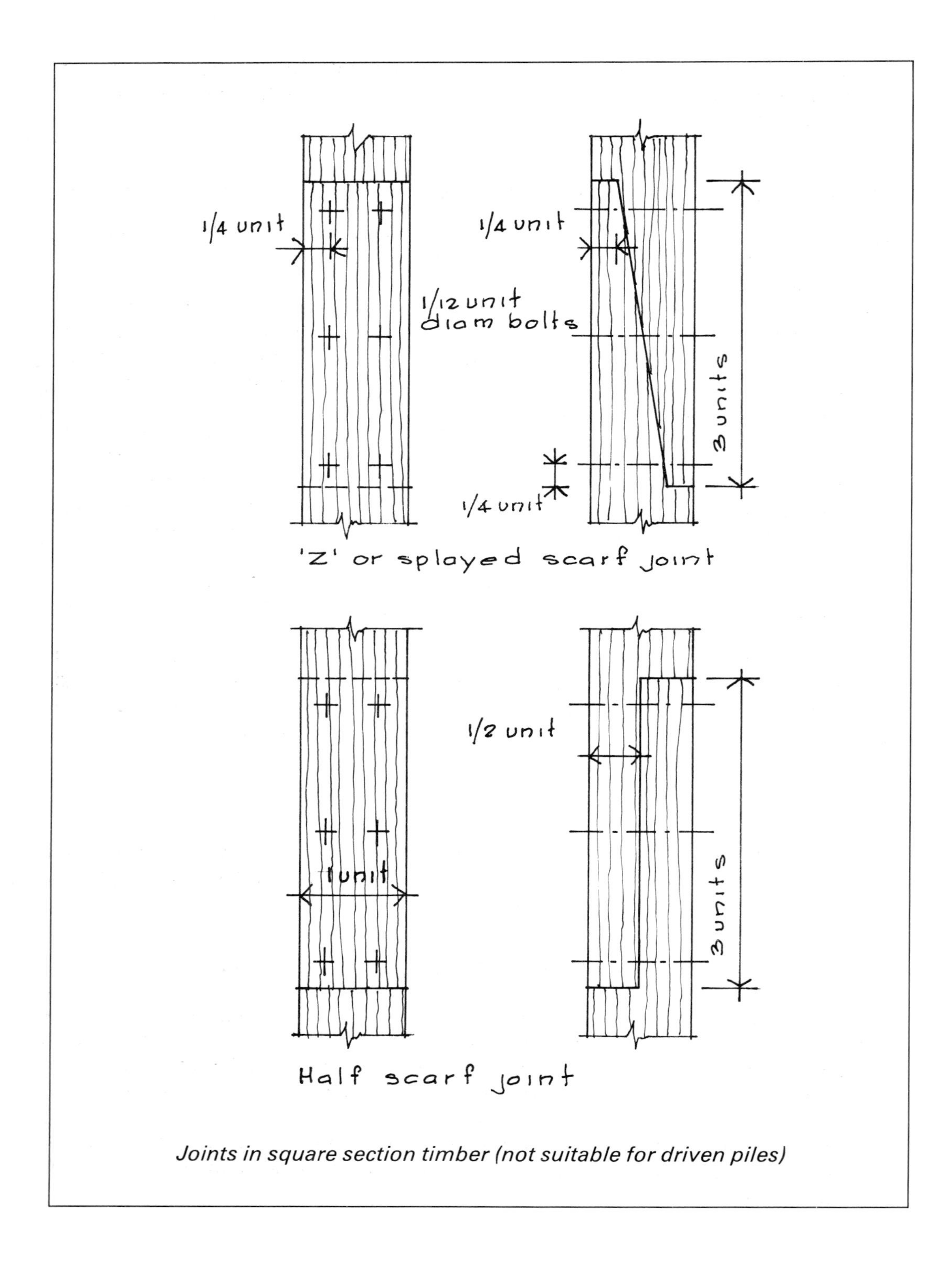

Joints in square section timber (not suitable for driven piles)

One can try water-jetting down the side of the pile and into the river bed, as far as the pile toe if possible. One resistance to overcome is the suction of the pile in the ground; jetting relieves this.

Piling hammers and extractors

Most types of hammers can be used for driving all kinds of piles. A drop hammer needs a pile frame with guides, and a winch to drive it. Most others are suspended from a crane with leaders attached.

The choice of hammer is usually a matter of experience, the type and weight of pile to drive, and the material to be penetrated. With a long, heavy or bulky pile it is often better to start with a modest hammer. It is more easily controlled when the pile has not reached sufficient penetration to support itself. A heavier hammer can then be employed to obtain the necessary penetration and/or set.

Vibrating extractors also relieve the suction and help to lubricate the friction between the ground and the pile or tubular pile. If all fails, the authorities will usually allow the piles to be cut off one metre below bed level or below any future dredging. Sometimes it is better, both economically and politically to get the authority to attempt to extract the pile(s). If the authority cannot pull it, one has a good case for cutting the pile off.

A timber pile can be pulled sideways in the hope it will break off one metre or more below bed level! This would save the cost of a diving team excavating around the pile either to cut it off or use explosives to sever it.

Steel sheet piling is usually extracted by an appropriate pneumatic extracting hammer, clasped to the top of the pile by special jaws. A crane is used to add uplift and to take the weight of the loosened pile, once free, and the extractor. Vibrating extractors are used as an alternative in suitable ground conditions.

Cofferdams

The design of cofferdams is for the expert. Above all they must be strong enough to withstand the pressure of water from outside. Generally they are constructed in interlocking steel sheet piling with steel walings.

There are several precautions to observe on site:-

(i) Before "closing" the last pile, all the internal support walings must be in place and fixed.

(ii) Always have relieving valves fitted at a low level and operable from the top level, to allow flooding of the dam to relieve external pressure if there is likelihood of collapse, or, if by chance the cofferdam is flooded before a falling tide, it can be relieved from internal pressure as the tide falls. Rarely is a cofferdam designed to take such reversed forces.

(iii) Make sure the designer has given a safe "cut off" in the ground. London clay is best. Beware of chalk – its fissures are treacherous and can lead to bad leaks, or even a "blow".

(iv) Make sure the labour force has adequate escape routes to the surface. Remember wooden ladders will float away if not held top and bottom.

(v) When driving sheet piling for a cofferdam, ensure that plenty of old grease is applied to the male and female joints. The labour force will hate doing this but the task is worthwhile. All cofferdam piling will leak somewhere, but this can be minimised by using a mixture of grease and boiler ash. Once the piles are driven and the cofferdam pumped out, leaks can be reduced or stopped by a few handfuls of ash being placed in the water immediately above the offending leak. Even as one watches it should stop the leak by bedding itself into the grease. This operation should be carried out at "slack" water, otherwise the ash is swept away.

(vi) Where a cofferdam is three sided and closes against a wall on its fourth side, difficulties may arise. The scope of these will depend on the alignment and condition of the wall and whether or not it has protruding footings. Beware of brick walls – great chunks can come away when lateral pressure is applied.

There are numerous means of plugging between the piling and the wall; for example, wooden wedges, lead wool, inflatable bags, etc. Do not forget that these must be placed from the outside, by divers if necessary, so that the water pressure will help them to stay in place. Any such plugging should be checked regularly, particularly if there has been a reverse head of water from inside.

(vii) A "blow" in a cofferdam is sudden and can cause panic.

Fendering

The intention of fendering is to protect a structure, if necessary at the expense of damage or loss to that fendering.

The earliest form of fendering was usually a softwood timber pile (or baulk) fixed to the structure immediately in front. Then came hardwood for greater strength and durability and then various hardwood rubbing pieces to protect the fender pile.

Fendering is in my opinion the unresolved problem in river civil engineering. Does one spend money on sophisticated fendering to save money in the design of the main structure or does one give more attention to flexibility in the structure and reduce the fendering?

To list but a few of the inventions introduced since the simple system referred to above, there have been or still exist:-

(a) **André Rubber Co.** – "Shear" and "Raykin" types and tubular rubber.

(b) **Gravity fender** – A very large concrete block slung on chains which in turn are fixed to the underside of the jetty.

(c) **Spring fender** – as at the entrance to Tilbury Docks. This is a concrete arm some 150 metres long and carried on steel section piles, which is designed to deflect.

(d) **Coir fender** – Made of rope as put over the side of a craft.

(e) **Japanese inflated fender** – Floating in front of the structure.

(f)	**Cambridge fender**	–	Mild steel bar, 150 mm diameter, which is gradually twisted by each contact of the vessel.
(g)	**Peine Pile**	–	A row of driven steel sheet piles or single piles stiffened up by steel joist or beam sections.
(h)	**Steel box piles or tubes**	–	Filled with concrete.
(j)	**Floating timber booms**	–	In front of simple vertical fenders.
(k)	**Larssen single pile**	–	Bolted to solid vertical face, pan outwards, and the cavity filled with concrete.
(l)	**Lord Fenders**	–	Arched rubber between fender pile and jetty deck.

All the above are in various sizes and held in position in various ways.

In considering fendering, the reader can only assume a craft will approach at an acceptable speed and angle and he should apply whichever fender fits the case. No fendering will withstand a vessel running amok.

This happened with the "Monte Ulia" at Coryton. This ship felled an entire jetty head and its superstructure, all of which disappeared underwater. The ship carried on through the jetty approach, dragging away a 600 mm crude oil line. The resulting oil slick caught fire and floated towards Canvey Island, burning two barges on the way.

Booms – floating

These are designed in timber to float in front of jetties, spanning the vertical fendering and thereby spreading the force of impact from vessels berthing alongside. The cross section of a boom must be wider than its height, otherwise it will try to float diagonally, which defeats most of its object. Alternatively, the boom can be round in section. Booms should be designed to resist impact, serving as a beam between the fenders.

The usual guides for booms are made of stud link chain, which is less likely to "kink" than other types. The chains pass through the boom at each end (through prepared and sleeved holes) and are secured at about deck level. A sinker of sufficient weight is required, resting on the river bed to keep the chains taut. If the water is more than 10 m deep at low water, it is better to fix the bottom end of the chain to a jetty pile about 10 m below low water to avoid slack which allows the boom to drift about.

Round booms, which are designed to revolve to avoid frictional wear from perpetual movement, have expensive swivels on their ends through which the chains pass. Wire does not provide a good alternative to chain, due to its comparative lack of weight and tendency to rust.

My experience has been that booms for commercial craft cause more damage to the jetty and themselves than if they are not used. Wear soon takes place on the softwood boom; this is caused by the vertical hardwood or steel rubbing piece on the jetty. A lateral blow from a craft on a worn jetty boom is likely to take off several rubbing pieces. An impact at one end of the boom will bring the force of the impact on to the end fender pile.

Unless the vertical fenders are supported above low water over their full length, they are in effect a beam supported at deck level and at low water. Alternatively they are a driven pile supported at deck level and river bed. The main disadvantage of a boom is that it transmits the impact into a knife edge blow on the fender instead of the shock being distributed by the craft's side.

Over many years booms were devised and found to be unsuitable, except perhaps the very modest ones to keep dinghies and small sailing craft off a mooring face or from falling between fender piles spaced wider than the length of the craft. I have experienced booms 26 m long by 800 mm wide and 400 mm deep, with 38 mm stud link chains, 3 tonne sinkers and mild steel connecting plates, jammed between a large tanker and a jetty face on a falling tide. This has left the boom some 6 m above water level, with the result that when the craft eased off, the boom fell to the water line causing extensive damage.

The Japanese make enormous inflated sausage shaped booms which have proved successful on jetties designed to receive them. These are a totally different conception to those described above.

Debris

It would seem that waterways are considered by the public to be a free dumping area. In spite of the efforts of river authorities, barge load upon barge load of debris is to be found in the water. The variety of material is incredible. Years ago, when the River Thames was busy, persons would gain a living just by retrieving floating rope, which was later sold ashore for paper making. I saw one such individual plying his trade only recently.

Polythene sheet is notorious for collecting at water intakes and possibly blocking pipes. It will also block cooling water intakes on motorboats with disastrous results. The flotsam is blown about by the wind and gets trapped in corners and amongst craft.

My attention was drawn to the matter of debris when my Company contracted to place booms across the Thames at the upper and lower limits of two separate power boat race courses to prevent debris entering the race areas; one in the County Hall area and the other off Battersea Park. As far as I am aware this had never been attempted before. The first effort consisted of 300 × 600 flat timber booms stretched across some 150 to 180 m width of river, with 150 mm square mesh reinforcement fixed vertically to one side – 1 metre below water and 150 mm above. The top of the boom was at water level. The sections of the boom were wired together in a continuous line and restraining wires were attached to conveniently placed bridge piers and ground moorings ashore.

The Port of London Authority would only allow the booms to be placed about two hours before the commencement of racing and to be removed immediately afterwards. We had only been permitted to rehearse with short lengths prior to the day.

The object was to catch the debris with the down river boom on the flood tide and, once the debris in the course area had cleared, some 1600 metres towards upriver, the upper boom would be "closed" ready for the ebb tide. Each boom was to have at least six attendant motorboats with six men in each to grab out the trapped debris by hand and throw it into a nearby barge. The whole operation was to prevent debris entering the course and damaging the racing boats.

I learned that:-

(i) the ferocity of the Thames when it is trapped is tremendous.
(ii) the build-up of debris was alarming and overwhelmed the boat crew's efforts.
(iii) large and small debris appeared to "see" the booms and were swept under the 1 metre deep mesh to be trapped by the eddy on the wrong side of the boom.
(iv) saturated and drowned rats, dogs, pigeons and chicken are not pleasant to handle.
(v) some nine racing boats were sunk by escaping timber baulks.

The experience could be enlarged upon!

When the second race meeting was announced a year later it was thought that lessons had been learnt. However, as mentioned above, this race was to be held in a different location.

Some time before the second meeting a Company offered the sponsors, without charge, inflatable booms with similar attachments to the timber ones. These were originally intended for oil slicks (i.e. they were without the mesh, etc.). My Company contracted to place them and supply the attendant collecting boats and crews, as before. The inflatables floated on the water. No deliberate sabotage was committed – unless the Thames wanted to show its contempt – but most of the debris seemed to have sharp edges and spikes that day and the sinking of punctured booms outweighed the casualties to the racing craft. On this occasion 48 hours was given to place the booms, provided a staggered gateway was left centrally to allow regular waterborne traffic to pass through. The P.L.A. required continuous lighting along the top of the booms, but the weight of the lamps and cable turned the booms "turtle" and this arrangement had to be abandoned.

These anecdotes are intended to show that it is impracticable to clear the river completely of floating debris. Remember also that submerged logs and debris roll along the river bed and cannot be seen from above.

Crane barges and pontoons

The fitting out of a crane barge or pontoon calls for a certain amount of design and even more common sense. Above all the craft needs to be as stable as possible. Working from a lively platform is impracticable, frightening and uncomfortable for the crew. Ballasting is necessary down in the bottom of the craft; concrete, or precast blocks or granite setts, etc., are suitable materials to lower the centre of gravity and metacentre. A transverse ballast wagon on track (well secured when not actually being moved) will help to trim the craft during various working loads.

Adequate wires and fairleads are essential, particularly if the craft is to be used in the tideway and/or is required to be held in one position. Pile frames, cranes, etc., must be held down with proper attention to safety factors. Avoid open holds which can flood with wave action. Coamings can prevent a good deal of water entering the hold.

Crane tests must be carried out to satisfy your Insurance Inspector and the Factory Inspector. Crane barges and pontoons can be self-propelled and, so far as I am aware, any person may navigate, though it is mandatory for lightermen to be aboard when towing takes place in the Thames and Medway.

Marine insurance is an expensive and intricate subject. The cost of premiums is some eight times that of a comparable "all risks" cover ashore. The reader should declare the plant on board and insure it separately from the craft.

If the craft is damaged by collision with other craft, by law the offending craft will only pay a lump sum based on its registered tonnage. A tug of 100 tonnes gross could sink the reader's craft, or cut it adrift with consequential damage to his vessel and other property, but the tug would only be liable for the first few thousand pounds.

In the event of the reader's barge or pontoon, whilst being towed by a hired tug, becoming damaged or causing damage to other craft or property, there is no redress from the tug skipper or company.

Should the reader's barge or pontoon sink, either under tow or at a mooring, the river authority should be informed immediately – unless the authority has phoned in the middle of the night with the news! The authority will take it on themselves to mark the wreck and salvage it without delay. The usual procedure is to lift the sunken craft sufficiently to beach it on the most convenient foreshore, so that the owner can carry out any necessary first aid, to refloat it and move it away for permanent repair. Quite often a sunken craft is badly damaged in the lifting operation, but the authority will not consider a claim for this.

The claim procedure is:-

(i) The marine and plant insurance should cover the cost of the primary salvage operation – i.e., the authority's charge, plus any other expenses.
(ii) The cost of the first aid operation should also be covered.
(iii) The lifting damage and permanent repairs are covered up to the insured value of the craft; likewise for the plant. If the value of the craft would be exceeded by the cost of repair then it is considered to be a "write off".

A salvage and repair, perhaps together with taking away damaged plant, can be a long job. It is necessary to negotiate with the underwriter's surveyor and keep him informed of (and in agreement with), the proposed procedure. These surveyors are experienced professional people and know every answer. Be wise and do not try to make false claims.

I found that these surveyors do not want to deny an owner any reasonable claim and have even been known to give sound advice. On the other hand, they have their clients to protect and will make frequent visits to the barge repair-yard to monitor costs and scrutinise everything put before them.

On hiring a craft all such things should be considered and the insurance payment and total loss value should be agreed before commencement of hire.

Taking craft out of a river calls for special insurance and Board of Trade Certificates. These Certificates are only granted once a craft is found to be seaworthy by a Board of Trade surveyor and the required fees have been paid. These arrangements can take a week or so over and above the time required to carry out any necessary work.

The uses of a crane-barge are endless, but however used, do have a foreman of outstanding quality and experience in the tideway.

It is common to have six moorings laid out when away from the shore. The usual pattern is a head and stern mooring and two breast-moorings on each side. Although anchors are usually employed at the ends, the secret of anchorages that do not drag is plenty of heavy chain attached to the anchor and laid on the river bed and, one hopes, in some mud. A tug will probably be required to "run" the anchors, chains and wires. Usually a vertical lift is needed from the craft to pick up the anchors although a tug will still be required (except for self-propelled craft) to take charge of the craft once the moorings are taken up. Make sure the moorings are long enough so that their catenary is little affected by the rise and fall of tide.

Avoid fixed moorings to a jetty etc becoming too taut on a falling tide; this may cause possible damage and breaking adrift. It is necessary to be constantly vigilant with moorings of all kinds. Even licensed lightermen and watermen can make bad mistakes.

Before such moorings are laid or craft moored in the open river, the authorities will require notice and a fee paid, in order that a "Notice to Mariners" can be issued, informing users of the river of the craft's position and moorings (See Page 49).

When working under bridges or near jetties, make sure the craft is not trapped under the soffit on a rising tide; otherwise something has to give. Most structures are designed to take downward forces, but not always upward. If the structure withstands the considerable uplift from a craft, then it may be swamped and sunk as a consequence – as was my crane-barge when trapped under Westminster Bridge!

Adequate cabin, toilet facilities and fresh water must be provided for the crew under the Health and Safety at Work Act. Sewage cannot be disposed of overside, even after maceration. A safety boat is essential as well as having a "jolly"-boat to ply between the working craft and shore.

Groynes

These can be found in the lower reaches of rivers and on many estuaries and coast lines. The object of the groynes is to prevent erosion by conserving and collecting beach material.

Groynes are constructed of timber – opepe being the most economical and suitable – with wooden piles and planks as cladding fixed with galvanised coach screws on the side of the prevailing wave action.

As a general rule the pile should be driven down twice as far as that part remaining out of the ground, which to be practicable should stand at a maximum height of about 2 to 2½ metres above the beach level. A post and "cage" type beacon is required as a warning fixed to the end pile. The beacon should be 3 metres above high water.

In constructing groynes towards the open sea there should be a safe haven provided for plant and machinery, preferably above high water, as storms are capable of causing serious damage to such equipment.

Tetrapods

A proven method of coast protection is to place at random sufficiently heavy and rugged concrete or stone blocks; these will break up wave action.

This somewhat haphazard method has been much improved upon by the introduction of the French patented invention of Tetrapods. These are cast in concrete within prepared steel shutters. These are hired from a Tetrapod-hiring agent. They are partly hinged so that before placing the concrete filling the opening side can be closed and fastened. There are various sizes to suit the area of coast line to be protected. A suitable number of shutters should be hired to allow for economical pouring to produce the right number of completed tetrapods. Suitable mould oil is also required.

After striking and curing completed Tetrapods can then be lifted and placed, again in a random fashion. Because of their shape (somewhat like two Isle of Man images entwined) they interlock readily. No reinforcement is used so that rapid construction is possible.

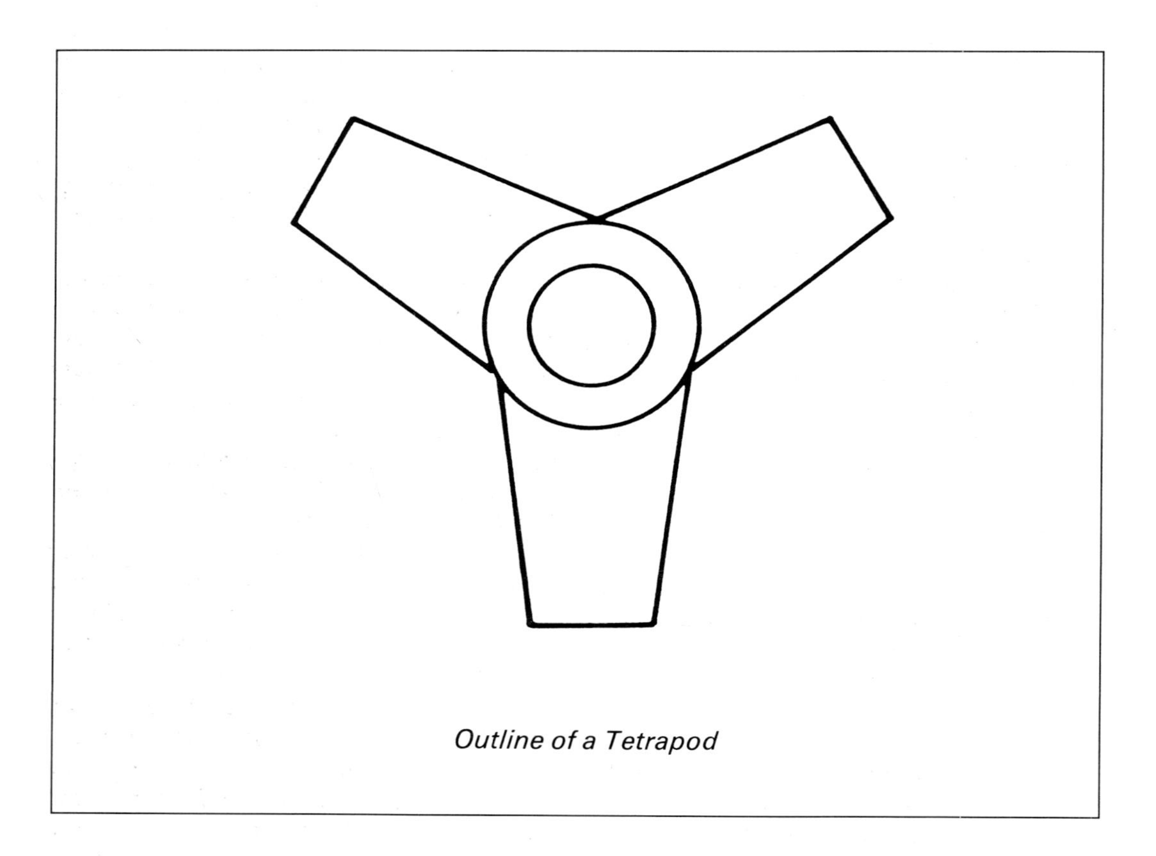

Outline of a Tetrapod

Jetty Construction

The main materials used in jetty construction are timber, reinforced concrete and steel – historically developed in that order. Pre-stressed concrete had a short life after being introduced in the 1950s. Following damage to a pre-stressed structure, repair is virtually impossible without reverting to new sections of traditional reinforced concrete; the jetty can thus become a hybrid.

Timber structures, up to 1950, were cheap to construct and easily repaired. They were resilient to impact from smaller craft, and were designed in the days when a 10,000 tonne merchant ship was considered to be large.

Traditional reinforced concrete piles are difficult to handle and easily cracked. Handling these over water, together with conveying and placing wet concrete, to complete the construction of the structure, is far more difficult and expensive than ashore. A good deal of temporary staging has to be constructed in advance and then later dismantled. This applies to cased piles, which need concrete and reinforcement conveyed to the site and placed. Owing to floatation they also have the disadvantage of carrying out a merry dance when being "pitched" into a "gate".

Soffits to low water slabs (diaphragms) are easily supported from piles, but trying to strike the shuttering (from underneath the completed slab) afterwards is not only very difficult but also expensive because the operation is governed by the level of low water. Working under the soffit, say from a raft or boat, can be particularly dangerous to any men being caught by an unexpected swell or wave action.

Steel structures are usually rigid and, as a consequence, require plenty of fendering. Steel piles may deflect slightly but soon take on a permanent 'S' shape at bed level if pushed over too far by lateral impact.

It is my view that the best arrangement is to construct a jetty as a means of access to and from the craft, with isolated strong points (such as breasting dolphins) just in advance of the jetty face line for craft to berth against. When a craft berths, very rarely does it fall alongside exactly parallel with a jetty face, but invariably a "point" contact is made, so a small but adequately strengthened area is desirable to take the first impact, before the craft finally breasts against all the strong points.

Mouchel and Partners have designed a number of dolphins constructed of large diameter steel piles with "loose fitted" concrete decks and diaphragms. In certain circumstances these dolphins can deflect 750 mm without damage.

Pipework, conveyors etc. on jetty heads and approaches are best kept well away from the berthing line in order to minimise the possibility of contact with craft and their superstructure. If such equipment is carried above the deck they are accessible for maintenance and inspection, but clutter the working area; if slung underneath, then a clear deck can be provided, but with more difficult access and maintenance.

Steel and timber constructions suffer from the effects of the "wind and water" line which is at and just below high water. At this level and below softwoods and elms are affected by rapid rotting, and steelwork with rusting. It is imperative not to mix steel and timber in a jetty's main construction. Due to necessary oversize drilling in the timber members the connecting bolts are never "solid" tight and this allows the timber to "work". This has the effect of the steel members sawing through connecting bolts.

Cathodic protection was introduced in the early 1950s. As the protection is intended to operate over a number of years, it was not possible for me to monitor the success of the system, but I have always had doubts of its efficiency for the reason that the most vulnerable part, being the "wind and water" area, is only protected apparently during the high water period.

Jetty conversion

There is little to say on this subject. It is my experience that a jetty designed to handle a particular commodity never seems to be 100% successful in handling something else, however much the structure is adapted.

Ladders and handgrips

Ladders to river structures are usually provided as a vertical access for personnel climbing to and from smaller craft coming or lying alongside. These ladders reaching from the lowest low water to deck level, are usually of mild galvanised steel and are fixed at these levels and two points between. Rungs should be about 250 mm centres, and the ladder at least 400 mm wide.

Ladders should be set within the line of the main fenders. If the main fenders are so far apart that the ladder is not readily protected by them, then additional fendering is advisable on each side of the ladder. These additional fenders also have the effect of allowing craft which may be too short in length to span the main fendering, to lie alongside the ladder without damage. Ladders should be parallel to the line of the river and not at right angles to the flow, as it is not practicable to berth and lie small craft athwart the tide.

Without galvanising or similar adequate protection, corrosion will soon occur at the "wind and water" level and render the ladder unsafe.

At deck level, a handgrip can be let into the deck underflush, or set on the deck itself, though the latter can be a hazard. The stringers of the ladder can be extended and forged into an inverted "U", about 450 mm high and fixed down to the deck, but this method can be a nuisance when handling rope over the side of the structure. However, rope guards can be fitted each side of the handgrips to prevent ropes fouling.

Dolphins

This is a technical word not usually found in ordinary dictionaries. Dolphins are independent piled structures rising to the usual jetty level, which fall into four main categories:- breasting, mooring, protective and marker, or any combination of these.

(a) **A traditional breasting dolphin** usually presents a vertical surface for ships to bear on or lean against, with raking piles on other sides. Usually there is a deck, with access ladder if there is a bollard fixed to it, and sometimes navigation lights.

The design can be in timber, steel pile and concrete deck or a circular ring of sheet piles tied across and the cavity filled. Generally there is a form of fendering on the side receiving the ship. The size and design of these will depend on the size of the craft to be accommodated. Timber can be suitable for a coaster, which would arrive unaided by tugs, provided the length of the piles out of the river bed is not too great and give the craft a soft, resilient cushion to fall against.

It is not unusual for a steel-piled dolphin to deflect 150 mm on impact and less rigid tubular-piled dolphins have been designed to deflect even 750 mm. The longer time taken in deflection helps to reduce the amount and increase effectiveness of the fendering. On those with so much deflection, the concrete deck is not rigidly fixed to the piles.

A breasting dolphin can include a mooring for craft alongside, but the forces have to be catered for in the design. It is better to use such a mooring for a for'ard or back "spring" rope, on a fairly flat pull, to prevent ranging of the craft. A short lead from the bows with a near vertical pull will not prevent the craft ranging or drifting off. By far the greatest danger is an unattended rope pulling the bollard out by the roots; or the rope breaking on a rising tide allowing the craft to drift off. Alternatively, the rope may go slack on a falling tide also allowing drift.

(b) **A mooring dolphin** is designed for the purpose its name suggests, usually with one or two bollards allowing for a pull from the directions envisaged. A circular dolphin is useful in that it can easily take pulls in any direction and can accommodate a craft rounding a point – say in a dock – and needing a fixture to hang on to. Fendering would be required on the side or sides of the dolphin used by the craft.

Bear in mind a loaded ship at low water presents a very different set of circumstances to an empty one at high water. 12 metres range in height for a modest craft would not be unusual, and a "supertanker" would offer vastly more.

(c) **A protective dolphin or cutwater** is usually found at the ends of a jetty head and is often pointed on the exposed part of the dolphin. Originally these were intended to deflect something like a drifting barge from getting under, or damaging, the end of the jetty head. Cutwaters are not considered as being particularly necessary nowadays. However, where protection is required from moving craft then a dolphin of three piles or more can be very useful.

(d) **A marker dolphin** (or even one marker pile) is intended to mark a submerged structure, such as the end of a pipe extending into the river, or the outriver corners of campsheeting. These markers are usually provided as a requirement of the Port Authority and act as a warning to passing craft of a hidden obstruction. Sometimes the authority will call for a notice board to be attached to the marker to illustrate its purpose. These are illuminated at night. In addition navigation lights are sometimes called for.

Wharf fronts

Wharf fronts originated from the sinking of a wooden barge inshore and filling it to act as a deck. Craft, such as sailing barges, would berth alongside at high water and later ground.

Timber piles and campsheeting (a term of Roman origin) were widely used from the 1900s and many such frontages are still in existence. These are usually suitable for a berth which "dries out", so that the "passive resistance" of the berth helps to maintain the fill behind. The bottom boards rarely go much below the berth level, because of the impracticality of excavating so low in the tidal range, or even below low water level.

Hardwood rubbing pieces are used to protect the driven piles, which in turn are tied back with mild steel rods, up to 50 mm diameter, to timber piles or concrete blocks. These anchorages are sited behind the natural angle of repose of the ground behind.

Should the berth scour, for any reason, leaching can occur below the bottom boards causing settlement at deck level.

Reinforced concrete piles and reinforced concrete slabs were introduced in the 1940–50s in order to provide a longer lasting material and to avoid timber licences, which were in force at the time. The design was practically identical to the timber structures.

As larger craft arrived and a greater depth of water was required, steel sheet piling was introduced with walings, ties and anchorages. The anchorages took the form of concrete blocks, or a more easily constructed longitudinal reinforced concrete beam. An alternative anchorage to these and the old timber piles, were pairs of short length, steel-sheet piles.

Fendering is not necessary, provided no overhanging cap is fitted, as the steel piling can resist normal wear and tear from most craft.

Many steel piled frontages have been constructed in advance of existing defective or less adequate frontages. In these cases, provided the existing anchorages are found to be adequate the tie rods can be extended and used again.

When new anchorages and ties are required, make sure there are no previous frontages behind the one being covered up, which may impede excavation and the positioning of new anchorages. Again the new anchorages should be sited behind the natural angle of repose. If this is not possible, due to site conditions, then "A" frame anchorages should be considered; these can be effective at a much closer position to the new frontage.

Where there are buildings to prevent the use of the anchorages referred to above, or the forces are too great, then, after soil analysis, ground anchors can be considered. These are bored for and placed from the outriver side of the frontage, usually by specialists. This method is very expensive in comparison with the former method, and also wayleaves may have to be considered.

Continuous reinforced concrete caps were introduced to the top of steel piled frontages because it was thought this would act as a waling and improve the appearance of the piling. However, these had to be fendered to prevent craft catching under the overhang; and the caps were often damaged by over-riding the fenders anyway. Reinforced concrete caps were soon dispensed with in design. An inverted mild steel channel, with little or no overhang, fixed with cleats and welding is recommended. This improves the finished line and prevents exposure of the upper ends of the piles to the elements, thereby preventing lamination.

In nearly all cases of encroachment the authority will require an application for approval. Any work within 15 metres of the flood defences must have the approval of the authority. Charges are usually made for encroachment as for a purchase of the land.

Open quays

Any structure built in rivers will affect the tidal flow locally, perhaps causing erosion, but more likely siltation. Open quays, which are now constructed of vertical piles and deck only, have the advantage of the least obstruction to the tide flow. Older designs required walings and bracings which cause considerable obstruction.

Moorings

A mooring can consist of a simple wooden post, a small steel ring secured to a structure or land, cast iron bollards, or triangular lay outs of chain on the river bed which, in turn, are secured there by cast steel screws.

Here are some types in common use:-

(a) **A driven pile**. The top extends beyond the deck level by about 1 metre and is fixed at deck level, in order to resist a lateral pull. Alternatively it may be fastened to a vertical post of timber, or steel tube and fixed at two or more levels.
(b) **A cast steel or iron "bit" bollard**, bolted down to an adequate concrete or similar base. The most popular is a patented "Bean" bollard which is sold with specified load capacities. These consist of two posts on a base, with protrusions on the inshore face under which the mooring eye of a rope is passed. They have the advantage of securing two ropes and releasing them independently.
(c) **Piled dolphins** may need to be constructed to position and secure a bollard. It seems ironical that it is not uncommon to construct such a dolphin costing some £100,000 or more, to secure a bollard (the purpose of the exercise) costing perhaps only £1,000. A more sophisticated type of bollard is the quick release type which allows watermen to release the heaviest of a ship's ropes while some weight is still applied on it.
(d) **Traveller**. A vertical bull head rail or mild steel round (some 75 mm diameter) held only at the top and bottom and set at about 100 mm from the face to which it is fixed. A mild steel ring, usually 250 mm internal diameter and of 50 mm diameter material, is passed over the rail and is free to slide up and down to receive a mooring rope. A light chain from deck level allows the ring to be lifted clear of the water in order to attach the rope.
(e) **Anchors**. A craft can provide its own mooring by using its own anchor. River charts will show the areas where this is not permitted. Sufficient room must be given to allow the craft to "swing" on the turn of the tide, not forgetting that at low water the radius will be longer. A common use of anchors is with piling pontoons where usually a "stream" anchor is placed up and down river and two "breast" anchors are lodged, one on each side. With the aid of winches on deck the pontoon can be heaved into position and held there.

The secret of successful anchorage is a long heavy chain from the anchor to the mooring wire. The chain needs to lie on the mud and act as an additional anchor. Anchor moorings should always be as long and flat as possible. The nearer to vertical they become, the more chance there is of lifting the anchor from the bed.

Since anchors for a pontoon are likely to be at least 3 tonnes in weight, a tug is usually required to take out the anchor, chain and wire from the winch all to be dropped in position. Lifting out an anchor will require a crane or piling winch sited vertically over the anchor. A tug is required to control the pontoon before the last anchor is lifted.
(f) **Screw moorings**. The Port Authorities are about the only people with suitable craft to wind screws into the bed and with the experience to know the depth required to put in the screw. Depth depends on the material forming the river bed and the size of the

craft to be moored. From the screw a mooring chain or pennant is fitted, which in turn is passed through and fixed to a buoy sufficiently large to take the weight of the chain.

If more than one screw is required, assistance can be obtained using two others splayed at about 60° in line with the pull and tied to the first screw by chain. Each of these could be assisted by two more. The material for such layouts is enormous in comparison with other moorings and the cost that much more. Screws, unlike anchors, can take a calculated pull vertically.

(g) **Public buoys**. These are provided by the authorities as temporary accommodation for craft, and are of sufficient capacity to hold any vessel capable of navigating that reach of river in which they are sited. Fees may be charged for such accommodation.

(h) **Trots of Buoys, or Barge Roads**. This means a series of buoys sited for a particular purpose – again refer to Chart layouts.

Trots usually consist of sinkers (or even screw anchors) placed in line on the river bed, with a chain secured between them at bed level. Into these longitudinal chains rings are built in at the required centres from which pennants and buoys are fixed. Do not allow the moored craft to sit on the sinkers at low water.

Trots are common for securing "fixed" barges for others to moor to – these are known as Barge Roads and they have screwed anchors. Less heavy systems accommodate private cruisers.

Generally, consideration must be given to all types of moorings in order to allow for the rise and fall of the tide, remembering that there is more slack in the moorings at low water which will encourage drifting. Conversely, a small craft can be pulled under by too short a mooring on a rising tide, or hung up on a falling one.

A single mooring, or even a head and stern mooring, will allow a craft to sheer about, in a figure of eight, in the tide flow.

A pontoon can be fixed to serve as a landing stage (as in a Barge Road) for craft to come alongside to moor. This has the advantage of constant levels between decks for the benefit of passengers, etc..

Safety chains

The purpose of safety chains is either to prevent persons falling off the edge of the deck of a structure or, should a person fall into the water, there are vertical chains suspended from the deck to grasp until help arrives.

Generally it is left to the opinion of the local Factory Inspector whether or not any chains should be fitted. If chains are required they may be placed at deck level or both deck and tidal areas.

If fitted to the deck, it is my opinion that the stanchions and chains should be set back about 1 metre, so that they do not foul the handling of ropes, etc., to bollards. The rope handlers can then walk in front of the chains, while still hanging on to them as a personal anchorage. Leaning over a slack chain can be dangerous; alternatively, if the rope handler takes down a section of chain for access then this defeats the original safety measure.

Rigid steel tubes can be used as stanchions and horizontal rail but is subject to damage and a person can be trapped by a swinging load.

Chains hung over the side to reach just below low water can either be hung vertically or, better still, hung in catenaries, with the tops at about 3 to 5 metre centres. These safety chains are intended for persons to grab if they fall into the water.

All chains are usually of 15 mm diameter, and galvanised. The overside chains can have hand holds in the form of about 100 mm diameter rings at about 1½ metre centres.

Safety boats around working craft or marine construction are obligatory.

Signals and markings

There is now an International Code for marking fairways and approaches to rivers, etc. The Thames and Medway Authorities had to reverse the colour of the buoys in the channels, from green to red and vice versa, and alter a number of markings required on craft. Literature on all conventional signals and markings can be obtained from the authorities.

Those affecting river construction are:-

(i) Working craft with moorings and/or anchors laid – shapes and diamonds hung in a particular pattern to denote if it is a dredger, or moorings extending out from a craft etc. These are replaced by coloured lights at night.

(ii) Divers – red flag by day – red light at night.

(iii) Generally, in addition, the International Code flags YZ – asking passing craft to "take it easy".

Before commencing work with such craft, the river authority should be notified in good time to obtain their permission. Notices to Mariners may be required to be issued in advance by the authority and this can take two weeks. There is also a charge for these.

A reference to the section "Dolphins" includes remarks on marker piles.

Berths

There are two types of berth: one on which a craft can ground during the low water periods; and the other for when there is sufficient water at all states of the tide to allow the craft to remain afloat or to require only mooring arrangements.

Berths which are exposed at low water, or "dry out", can be visually inspected or surveyed by instrument to determine their suitability. Such a berth should slope slightly towards the river to allow mud to run off, but it need not be level in its length. It is imperative, however, that it is of an even plane. Any twist or undulations will affect the craft grounding on it, with possible permanent damage to plating, welds or rivets.

The best material for establishing a berth is "berth making" chalk. Small fine chalk is useless. Once disturbed underwater, say by propeller action, chalk is rarely re-usable.

If the "shoulder" of the chalk on the outside is, for example, 750 mm to 1 metre high, it can be supported by lead slag or Gabions. A more positive support is timber or steel piled campsheeting.

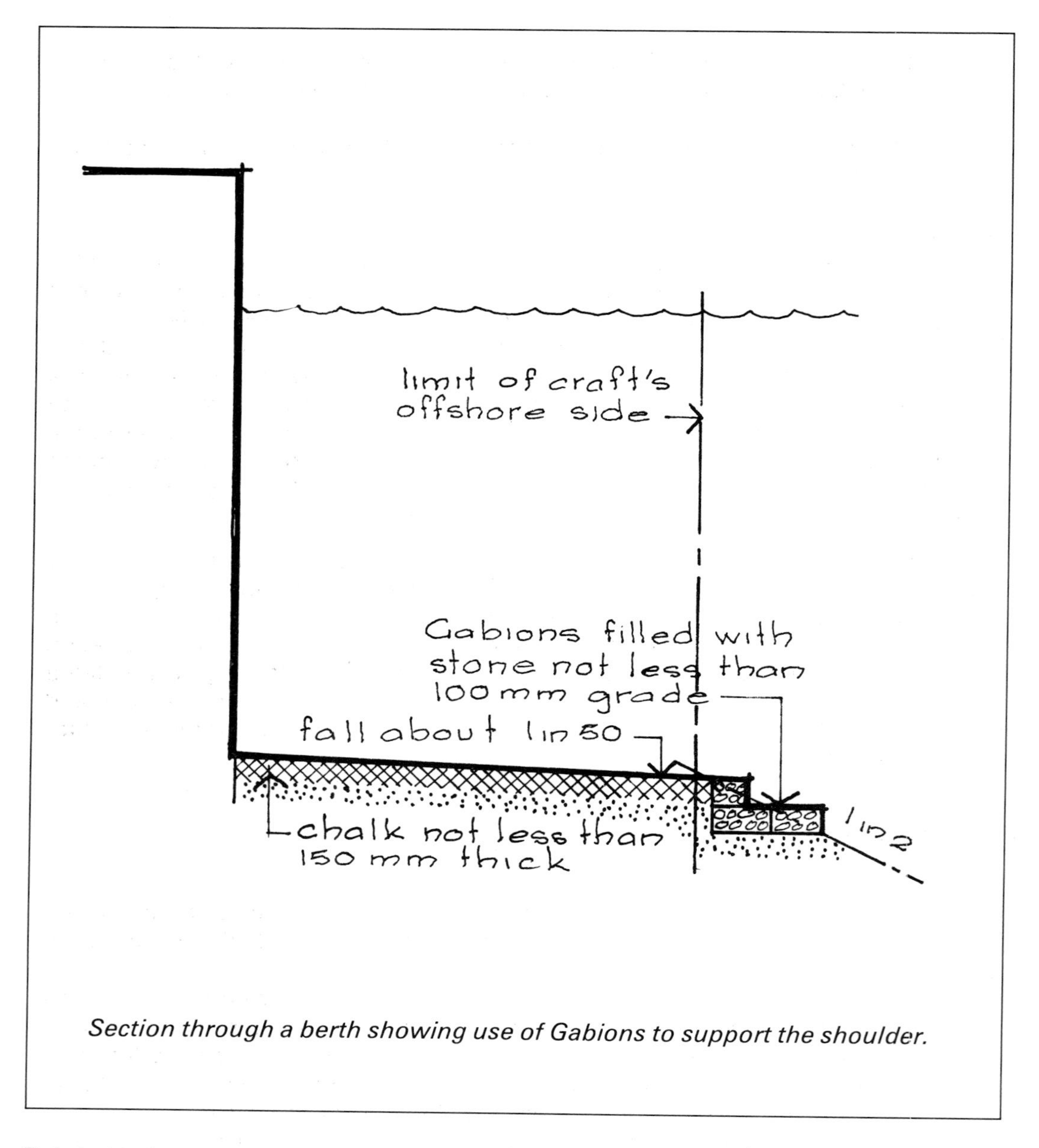

Section through a berth showing use of Gabions to support the shoulder.

Points to observe:

(i) Do not have a soft muddy berth, as it will tend to hold down a craft on a rising tide.

(ii) Do not berth different lengths of craft on a muddy berth. A larger craft will bridge the hole left by the smaller craft and perhaps break its back.

(iii) Do not let an impatient skipper try to arrive at or leave a berth without at least 1 metre of water under the craft's propeller; even then there can easily be damage to the berth by propeller action.

(iv) Do not let wharf owners invite ships to berth which will never have the minimum clearance. Owners always want to have ships just a bit bigger (and deeper) than before.
(v) Do not have a concrete berth. Hard debris will not be pushed into the berth (as with a softer berth) and the craft's bottom may be dented or holed.

On the other hand:
(i) Do survey berths regularly, before and after berthing, as craft owners are on the lookout for an excuse to claim bottom damage caused elsewhere or by previous negligence on their part. Bottom damage claims can be very expensive.
(ii) Look out for the skipper who, on berthing his craft, goes full astern to halt the craft's forward movement or to drop astern. Invariably this practice will gouge a hole in the best of berths and throw up a heap of debris on which the craft will sit at low water and receive bottom damage. The skipper will then claim a foul berth. If, before the craft arrived, the berth was surveyed, found to be safe and a record kept, then the wharf owner will not only be able to refute the skipper's claim but will also be able to claim against the ship for repairs to the berth.

A sophisticated berth can consist of driven piles and bearers, but it is really only practicable for smaller flat bottomed craft, with the bearers under the strong points.

When dredging for a new berth a licence from the river authority is required to remove material (and dispose of it above high water) down to the required depth. Once a level is established, further maintenance involving clearing accumulated material, such as mud and silt, must be applied for to the authority. This work should be described as "removing" accumulated material, *not* dredging, otherwise another dredging fee will be charged.

Brows and pontoons

A – Brows

This is a very common means of access from a fixed point ashore to a pontoon further out in the river. Usually the pontoon is sited to allow craft to berth alongside it at all states of the tide.

The brow consists of a hinged bridge, suitable for anything from pedestrians to vehicles etc., fixed at the shore end (clear of any flood prevention level) and resting with slides or wheels on the flat deck of the pontoon. To suit the Thames and Medway tidal ranges, a brow of about 25 m to 30 m long is best; with this length the gradients are within acceptable limits. Other rivers will vary in range of tide levels.

The sliding supports will move about 750 mm across the deck of the pontoon, owing to the rise and fall of the tide. For the sake of stability they should be central to the width and preferably central to the length of the pontoon. The deck should be non-slip to accommodate the traffic envisaged, with toe holds for pedestrians across part of its width. Handrails should also be fitted. The slides or wheels should be contained by fixed mild steel channels to prevent the brow "walking" about the deck. If the pontoon is even slightly out of level along its length, the brow without restraining channels will "walk" downhill during its constant movement.

Where extra length is required, rather than have a long brow it is better to introduce fixed bridge links from the shore on piled dolphin supports, with the hinges fixed to the dolphin further outriver.

In designing the "brow to pontoon" connection there is a certain amount of trial and error at the drawing stage. For instance, avoid the inshore edge of the pontoon passing the brow at high tides so that the slides or wheels are lifted off the deck. Always consider the highest recorded tide (obtainable from charts) and lowest spring tide. The brow can drop off the inshore edge, or be so steep at low water that it will "pin down" the pontoon on the rising tide with disastrous results.

Brows can reach direct on to a fixed craft; for example with the "Tattershall Castle" on the Thames Embankment and HMS "Belfast" in the Upper Pool of the Thames (see notes under "Pontoons" regarding craft "ranging" Page 56).

Brows can also be set parallel to river, as at the former P & O installation just below Tower Bridge. Here the brow was positioned over the length of the pontoon and it was necessary to have the hinged end high enough to avoid the rising pontoon being trapped under that fixing.

B – Pontoons

These can be of any size to suit the smallest craft or the larger liners (as at Tilbury Landing Stage).

It is best to control the position of the pontoon by means of a dolphin at the up and down river ends thereby leaving the inshore and outriver faces free for berthing. Alternatively, dolphins can be sited on the inshore sides.

Pontoons can be constructed from anything that floats. The best for the Thames and Medway is from steel plate built into water-tight compartments so that if one part is flooded the pontoon will remain afloat. Obviously the pontoon must be capable of taking the weight of the brow and any other deck loading and still leave a convenient freeboard. Steel will dent from impact without (perhaps) causing undue concern, whereas concrete is more likely to fail and is more difficult to repair. Polystyrene blocks, suitably protected, would serve very well for small craft. A Thames lighter is suitable but needs ballasting and decking over.

The deck level is arranged to best suit the deck of the craft coming alongside, but bear in mind that the pontoon deck level must be such as to avoid "heeling" of the brow (explained under "Brows").

Pontoons should not be less than 2.500 m wide or they become unstable for pedestrians. Of course they must also be wide enough to accommodate the "travel" of the brow (explained under Brows) and still allow space for pedestrians and/or vehicles to leave and turn in safety. Handrails should be fitted wherever practicable. A raised platform on the deck of the pontoon with steps in the direction of the passengers' walk can be introduced. This modification reduces the slope on the brow at low water periods.

The purpose of the dolphins is to restrain the pontoon in position on plan, and at the same time allow the pontoon to rise and fall with the tide. A "horn" at each end of the pontoon fits between two vertical guides on each of the working faces of the dolphins. The horn is usually made from steel 300 × 300 or 350 × 350 mm section and protrudes beyond the end of the pontoon and into the vertical guides. If timber guides are used,

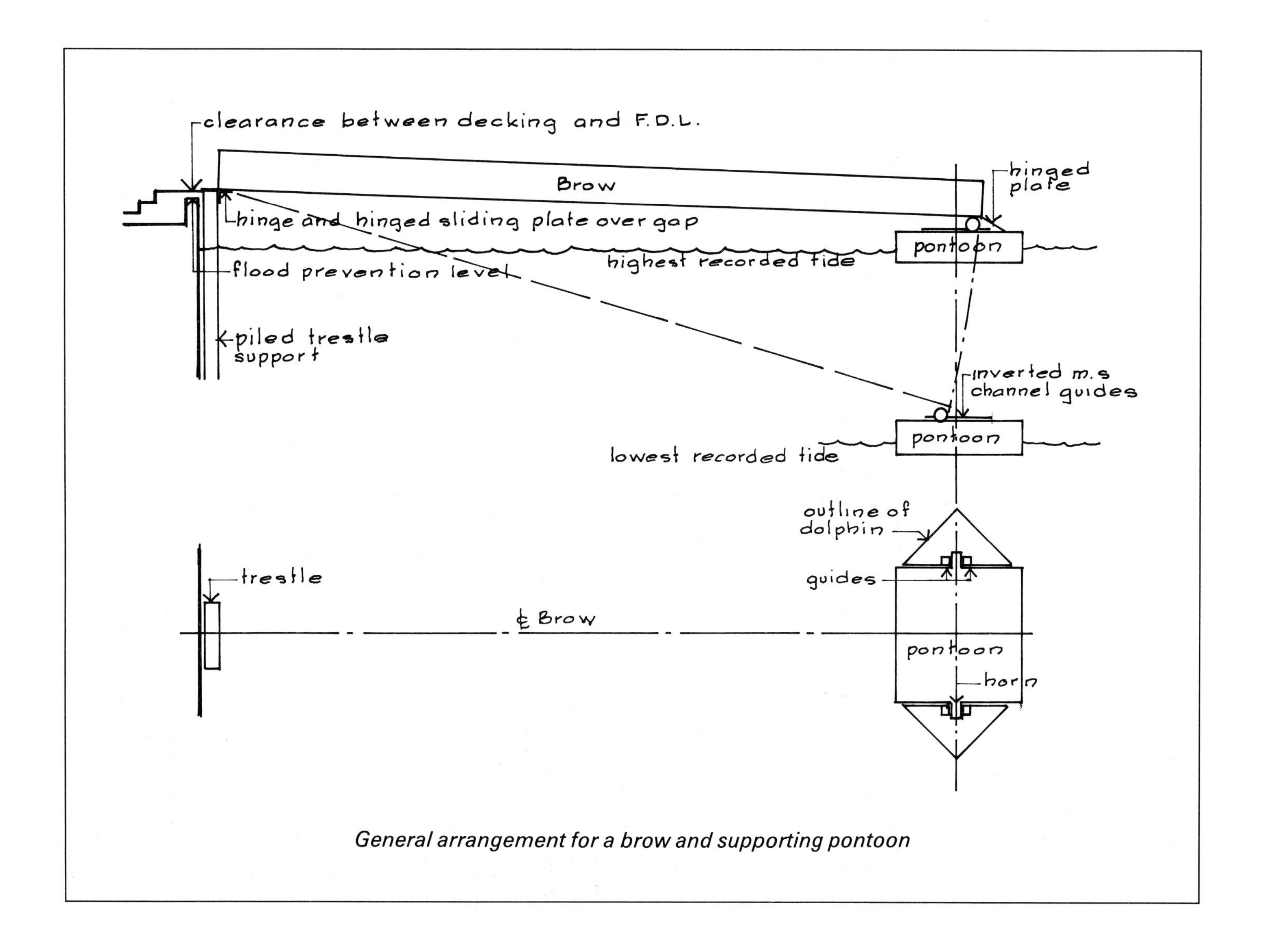

General arrangement for a brow and supporting pontoon

then wearing-plates, approximately 12 mm thick, should be fitted to the inner and upper, or lower, faces with countersunk bolts – not countersunk spikes as they shake loose and foul the horn's vertical movement.

The dolphins should be driven to reasonable tolerances. The vertical guides must be fixed exactly vertical, parallel to each other and square to the pontoon. The horn must not foul any walings, etc., in the dolphin but leave a clear passageway between the guides.

The top of the guides should be high enough to allow for the highest recorded tide, plus the freeboard of the horn, plus any swell, plus a bit to spare!

A tolerance of 12 mm each side of the horn and 35 mm between the wearing face of the guide and the pontoon at each end is the optimum. Any departure from these dimensions will cause too much movement of the pontoon or jamming of the horn. The pontoon is never absolutely still for a minute of the day or night, every day of the year. Driven piles as guides are not recommended because the accuracy required is rarely achieved in the driving.

The connection between the horn and the deck must be sufficient to transmit the live load of the pontoon and the berthing-loads to the guides.

The authorities will require "ground" chains to be fitted to the pontoon as an insurance against the pontoon breaking free and drifting into the main river (see section under ''Ground Moorings'' Page 49). If a craft is to lie permanently alongside, it must also have ground chains and preferably moorings to bollards on the pontoon. There has been much debate by designers on the loads and the effects on a pontoon by tidal flow, and as far as I am aware, so far unresolved.

The dolphins must be a rigid structure; any flexing may allow the pontoon to break free. Timber dolphins soon become flexible due to the bolted connections becoming slack. For easier berthing it is best to keep the outriver face of the dolphin in line with, or behind, the pontoon-face. It is also easier to allow a fixed craft to lie alongside the pontoon.

If a brow is to land directly on to a craft, the craft must be restricted in its movement except, of course, for the vertical rise and fall with the tide. Ground chains, or wires, fore and aft may be a safe mooring and satisfy mariners (and even the river authorities) but they will allow the craft to range in a "figure of eight" – it is a natural phenomenon. The placement of long breast wires towards the shore (at near right angles) from the fore and aft quarters, is the least that should be done to restrain this movement. "Breasts" laid to outriver, as well, is the real answer. A pontoon relying on ground-moorings will also perform in the same way.

Before breast wires were fitted to HMS "Belfast", she ranged 2.500 metres in all directions with the result that the 30 metre long, 10 tonne brow appeared to charge across the deck at an alarming speed, threatening to cut off the feet of those in its path and to remove the skylight to the Admiral's Pantry! Fortunately, glacier-bearings had been fitted to the hinge-end of the brow to allow for ranging, but so much movement at the deck end was not compatible with the safety of the visiting public.

The "Tattershall Castle" is "fixed" by hammer head horns, welded to its inshore side, which pass up and down between two pairs of driven piles acting as dolphins and guides. In this way there are three wearing faces on each pile.

The use of driven piles as guides with the "Tattershall Castle" contradicts my earlier statements that dolphins must be rigid, and driven piles should not be used as guides. Certainly I had grave misgivings about the success of this assembly. However, it has been installed for some years and it can only be assumed that, as the piles are "free standing" and greenheart, there is sufficient resilience to take care of the inaccuracies between the pairs of piles.

When a craft is fixed by its sides to dolphins there is an effect of jolting when, usually as a result of passing vessels, the craft strikes the guide. This can make normal foot passage unsteady and be embarrassing when taking soup on a restaurant boat!

Summary on Brows and Pontoons

For the benefit of any practical solution, it is well worth visiting as many brow and pontoon installations as possible on the riverside. There are all shapes and sizes. The writer has been involved with nearly all of those on the Thames and Medway; each has its particular functions and illustrates the development of techniques over some fifty years.

Beware of an updraught on open stairways from pontoons. Within hours of Her Majesty the Queen (now the Queen Mother) arriving by water to open Coryton Refinery in 1951, my carefully thought out open stairway brow, designed specially for this "one off" low water arrival, developed an alarming updraught through the treads. Sheets of plywood were hurriedly fixed to the underside to avoid embarrassment all round!

Causeways and stairs

These are to be found along both banks of the Thames and Medway. Virtually, they are footpaths or means of access from the shore line down to the lowest of the low waters. The slope is as gentle as the traffic and local conditions permit. In the main they are constructed at right angles to the river, carried on piles over the foreshore and decked over their width. Others are constructed in masonry, or concrete, built on the foreshore. Toe rails can be fitted transversely for pedestrians and a longitudinal kerb on each side is recommended for safety reasons.

The prime object of causeways is to allow craft, usually motor launches, to come alongside so that, without having to climb ladders, passengers can step ashore at any state of the tide, there being always sufficient water somewhere along the causeway's length for the craft.

Some of these structures on the Thames date back to the reign of Charles II and these historical ones are jealously guarded by the licensed watermen of the Thames. They are public rights of way and must be protected when considering development in the area. In recent years doors have been fitted on many at their inshore end, as a safety measure to prevent children gaining access. Keys are usually held by a neighbouring tenant.

Note: "Foreshore" is the area exposed by the tidal range between High Water Ordinary Spring Tides (H.W.O.S.) and low water. The "river bed" is that which is not uncovered by the tide. Make sure you use the right expression when talking to river personnel!

Slipways

The chances of a ship's slipway being required in a river nowadays is remote. However, slipways for sailing dinghies and the like are still being constructed, mainly by clubs and commercial marinas.

These are usually of a timber piled structure with about a 1 in 10 (10%) slope starting above high water level and extending to such a level as to allow craft to float off their trailer when the tide level is sufficiently high. The length can vary from that needed to reach deep water at low water level, to a short length to accommodate only high water.

The average width is 3 to 4 metres and a longitudinal kerb on each side, some 75 to 150 mm high should be fitted to resist trailer wheels going over the side. Toe-rails down the centre for pedestrians are advisable. The advisable loading from the trailer and boat onto the boarded deck is 500 to 1,000 kg.

Whilst one should design to meet the obvious hazard with "live" loads, wave action, causing a "slapping" of water under the decking, may cause the planks to be torn off by uplift. This damage can extend to the deck-bearers and capsills. A 25 mm spacing between the deck planks will relieve the force of the wave action but bolting or strapping down of every member is advisable.

Beware craft becoming caught under the overhang of the decking, should they approach or lie alongside the sloping deck. Rubbing boards or fenders are desirable, even in wave conditions of only a few centimetres or more.

Tide flaps

These are hung over the ends of pipes which discharge into the tidal range, to prevent water from the tide flowing up the pipe.

Basic types of flaps are hinged on a single pin, but the double hung, although more expensive, are less likely to stick open. Small diameter flaps are prone to sticking open because they are not heavy enough to overcome the buildup of silt etc., around the pin. A light chain, operated from deck level, will help to keep them free.

Debris can float between the flap and pipe end, if the discharge is slight, and prevent the flap from closing, so that on a rising tide it will remain jammed open by the head of water on the outside. A grille in front of the flaps is an advantage if the site allows.

Where roof drainage or surface water is passed from behind (or through the deck of) a flood defence to discharge into the tidal area, this would infringe the Flood Prevention Regulations (referred to in the section "Regulations" Page 4) if water backed-up through the pipe at high water. The backing-up can be prevented by inserting a special 's' or 'p' trap into the system, with an inflated ball or cistern ball in the trap. This has the effect of passing the water into the river when the tide level permits, but seals off the return flow against a prepared face. The system just described does away with tide flaps in these special circumstances. The better tide flaps have gun metal closing faces.

Water intakes

Intakes of cooling waters should be kept well away from the return discharge of warmer water. The outer end of the intake should always be below the lowest low water.

Intakes attract silt and debris into the system if grilles are not fitted over. Plastic sheet, however, will adhere to these grilles so they need constant attention.

Most docks require intakes to provide water from the river, during the high water period. This will maintain the impounded level inside the dock. The intake-pipe can be as large as 2.100 metres in diameter, and there may perhaps be three separate intakes together with a large pump for each pipe. Intakes to power stations and refineries are often multiples of 2 metre diameter pipes.

Silt and trash are a difficult problem to overcome. Trash screens need constant attention to keep them clear is one answer.

Lock gates

Lock gates on smaller rivers are usually formed in pairs. These are generally of greenheart or steel construction, hand-operated by using a long timber arm fixed along the top of each gate. Hand-operated paddles or sluices are raised and lowered on a cog wheel and rack principle.

As in all leaf-type gates, there is a vertical pin of about 50 mm diameter let into the base of the gate and into the sill on which the gate hinges. The vertical post at the hinge side is called a quoin. The top of the quoin is held back by a simple "U" shaped mild steel strap which extends back to an anchorage with thread and nut adjustment. It is essential that the rounded vertical face of the quoin fits exactly to the lock wall face, which is usually dressed masonry. A badly fitting quoin will leak water.

The pressure of water against the gate should help to close the quoin tighter. The other two points which should fit well are the sill and the mitred joint, where a pair of gates meet.

The sill is usually an upstand of horizontal timber some 150 to 300 mm high, set at the same angle as the closed gates. The sill is secured to a toe of concrete or sheet piling, which in turn is designed to resist the force against the sill when the gates are closed against a head of water.

Walkways, with handrails attached, are provided on top of the gates as access to the sluice-operating gear and to pass from one side of the lock to the other, over which-ever pair of gates is in the closed position.

On the lowest side of a lock there is generally a form of protection to the river bank – either sheet piling or stone-pitching. The repeated release of water through the lower gates can cause considerable scour. In addition, an apron of concrete or pitching is laid on the river bed, extending downriver to prevent scour below the sill.

Those few gates to be found in the lower tidal reaches of rivers are operated hydraulically. The most common of these are to be found at the entrances to main docks.

Dock entrance gates are made of steel and are in pairs, each single gate some 12 metres wide and 10 metres deep, or even larger.

Whilst the detail of the quoin, sill and mitre is similar to that referred to earlier, dock gates are much more sophisticated. The gate and sluice operations are by hydraulic systems, with back-up sub-stations and pumping arrangements. Hydraulic piping and electric cables are laid under the lock to facilitate the alternative pumping arrangements to be operated from either power house.

The gates are double-skinned, with chambers which are flooded to correct levels so that the gate is ballasted (to a plumb position) to prevent flotation, yet able to relieve the enormous weight of the gate from the gate's anchor points.

To remove the gates (i.e. one at a time) the top-anchorage is released as the gate's chambers are pumped out and the gate floats off the quoin pin. With help from suitable craft the gate is allowed to rise and lie flat on the water for towing away.

Silt can collect internally in the bases of the gates overloading the quoin anchorages and causing the bottom of the gates to scrape on the sill. Corrosion can occur to the inside of the gates, so manholes and vertical access shafts are provided within their thicknesses to allow inspection by divers. Greatest care should be taken during these inspections, as divers' equipment can get entangled and there is always a danger of gas above water level. The gate operating machinery should be immobilised during these inspections to minimise any possibility of accidents.

Restaurant Boats

If one is asked to install a restaurant boat, there are a number of items to consider which are not usually budgeted for by the owner. Some are statutory requirements, while others are a matter of common sense.

Invariably the engine has been removed, to afford greater space, leaving the craft like a cork on water. To regain stability the lost weight must be replaced (in the bilges if possible) with concrete, or better still, portable concrete blocks or granite setts.

In determining stability, one must consider such possibilities as, for example, all the customers and staff rushing to one side of the decks to see a person drowning or a collision, or, perhaps, suddenly evacuating the boat in the event of a fire. It is my opinion that one should consult a marine architect or surveyor at an early stage to advise on stability under these unusual circumstances.

Sewage cannot be discharged into the river. It is required that the macerating tank's contents be pumped out by a special craft, a visiting lorry tanker or pumped ashore by its own pumps. Usually a craft will pump out into a permanent main sewer or manhole nearby. Other services are usually required, such as telephone, gas, water, and a suitable electricity supply with switch gear aboard. Because all these services are likely to come from the shore, one must introduce flexible sections of pipe, etc. to each to allow for the rise and fall of the tide.

Again, usually, the access and egress to the boat is by means of a hinged-brow with the services slung alongside or under it.

In order to accommodate fire regulations and the comfort of the passengers, the maximum slope of the brow should not exceed 1 in 9, and the width must be adequate for panic evacuation. Often the Fire Officer will require a second brow, if only for the evacuation of passengers and/or galley staff. Completely enclosed brows are not usually acceptable – they can act as a "chimney" to the fire, or fill with smoke, etc.

Usually the owner will make himself responsible for the modifications inside the craft to comply with fire regulations.

It is my personal experience that a restaurant boat will be more suitably sited by being held with dolphins and horns, fore and aft, rather than alongside. If held at the side, there is a tendency for the craft to bump against the guides, which is disconcerting when taking a spoonful of soup!

Do not be carried away by the romance of a Thames sailing barge. Regulations, headroom and the liveliness of the craft are not for those without "sea legs".

Dock walls

An astute reader may consider dock walls not part of the tidal river. This is mainly true, but most dock walls extend beyond the lock gates and into the tidal area outside the impounded water and dock.

However, an Engineer engaged on river work is likely to be faced with work within a dock, so this section is included as a bonus of information.

Records of the dock walls in the London area are kept by the P.L.A. at Tilbury or by the London Docks Development Corporation at North Woolwich. M.P.A. have records for their walls at Sheerness. Other authorities will have records of their particular area or localities.

Surprisingly, other records are far from complete although it should be remembered that most of the walls were built before the authorities were inaugurated.

Most of the older walls are of "gravity" construction in brick, with a second brick wall built further back. Both walls are tied together and a filling placed between them. A good deal of damage has occurred below low water over the years and repair is costly. If the impounded water was ever drained off there is every possibility that the walls would collapse, owing to the hydrostatic head or other forces behind them.

A common method of gaining access to the wall below water level is by use of a "limpet", which is suspended in front of the wall and the water then pumped out. The effect is to force the limpet against the wall. It has to be heavy enough to resist uplift, unless adequate fixings are made to a sound wall.

A "limpet" is constructed of three sides and a bottom, and must be designed to withstand the forces from outside as well as scaled on the three connecting edges. If the wall is concave, the limpet needs to be shaped accordingly. Similar difficulties are presented in sealing a limpet as with steel sheet piling, referred to under "Cofferdams" (see Page 37). Before a limpet is considered an underwater inspection to assess the condition and profile of the wall is recommended.

Glossary of terms used in the text

Belly. Bend or catenary in a vertically inclined rope or wire.
Bent. A row of piles across the head of a jetty.
Blow. The sudden welling up of water and soil from inside a cofferdam.
Boom. A floating protective fender or device for containing oil spillage and floating debris.
Bottom damage. Indentations in the underside of the hull of a craft.
Bore. A tidal wave occurring at the commencement of a rising tide.
Bow wave. Water pushed up in front of a moving craft and fanning out as a wave.
Bracing. A diagonal member connecting vertical members.
Bucket dredger. A floating pontoon excavator involving a chain of digging buckets.
Cathead. The top of a piling frame or crane including the wheels.
Closing speed. The rate that a craft moves at 90 degrees to a berthing face.
Closing pile. The last (interlocking) pile in a cofferdam.
Coaster. A small ship working along coast-lines.
Cold work. Involves work with low heat sparks say from metal tools.
Creep. The tendency for the tops of steel sheet piles to move towards the direction of driving, giving the shape of a fan.
Datum. An arbitrary level from which tide levels can be related.
Daywork. Work carried out and paid for at an hourly or weekly rate.
Dip. The act of taking a particular sounding.
Displacement. The actual weight of water displaced by a craft, measured in tonnes.
Distance line. A light rope or wire marked off in predetermined spacing, usually for taking soundings.
Downriver. Towards a river's estuary.
Dries out. The act of the tidal water leaving a foreshore.
Ebb tide. A falling tide.
Fendering. Protection to the face of a marine structure.
Flood tide. A rising tide.
Gate. A means by which a pile may be held until it is sufficiently driven to be self supporting.
Guides. Pairs of vertical members containing the movement of the horn of a pontoon in a vertical direction.

High water. The highest level of a tide at a particular time.
Hopper. A barge to carry dredged material.
Hopper, calibrated. As above but with authorised markings on its hull to show the amount of material being carried.
Hopper, split. As above but hinges along its deck to discharge material carried through its underside.
Horn. A short projection from a pontoon to fit between pairs of vertical guides (see guides).
Hot work. Work involving arc or gas welding or oxy-acetylene.
Kickers. Usually timber struts laid on the ground to prevent objects closing on one another.
Ladder dredger. A dredger operating on an endless chain of digging buckets.
Low water. The lowest level of a tide at a particular time.
Metacentric height. The distance between the centre of gravity and centre of buoyancy of a craft.
Navigable channel. A prescribed route for ships to negotiate.
Pilot. A qualified person with local knowledge of tidal-flows and depths of water, who advises the master of a craft on the best route and manoeuvres to take.
Pitching a pile. Picking up and placing a pile in position before driving.
Pile shoe (diamond shape).) A pointed cast steel point on the driven end of a timber pile.
Pile shoe (sheeter). A flat shaped steel end to a timber pile which causes the pile to keep close to a previously driven pile.
Reach. A stretch of river quite often individually named.
Refusal. The point where a pile cannot be driven further.
Set. The number of blows from a piling hammer on a pile compared with the penetration at the final stage of driving.
Shot line. A weight (approx. 20 kg.) with light weight rope attached to a secure position near the waterline. The weight rests on the river-bed allowing the rope to serve as a guide to a diver to reach the required position of work.
Shoulder. The outer corner of a berth where it joins the original contours.
Sinker. A heavy weight to act as secure point on the river bed.
Slack water. When there is no tide flow, such as at high or low tide level.
Sounding. Measuring the depth of water over a particular point or area.
Steep approach. A craft approaching a structure at too acute an angle.
Steep sea. Waves which are more vertical than normal due to a wind opposing the tide flow.
Super tanker. An oil tanker ship within the top range of size.
Tidal datum. See datum.
Tide tables. A calendar of predicted heights and times of tables.
Tideway. Usually the area of a river used by craft.
Tidework. Work which is carried out at a time to suit particular levels of water in the tide range.
Top work. Work at or above high water level.
Trip. An obstacle to pedestrians.
Trot. A series of buoyed moorings in a line.

Turn turtle. To turn upside down, or invert.
Upriver. The direction away from a river estuary.
Waling. A horizontal member connecting piles or other vertical members together.
Walk. To travel sideways owing to constant eccentric longitudinal movement.
Welled up. To boil up in a river bed or cofferdam.
Wind and water line. The vertical height or length between the Spring and neap high water levels where wetting and drying by sun and wind occurs.

Bibliography

Tetrapods

Dock and Harbour Authority. London October 1958, October 1962, June 1987

Mooring craft

Oil Companies International Marine Forum. Guidelines and recommendations for the safe mooring of large ships at piers and sea islands. Witherby & Co Ltd. London -----.

Terado beetle, fendering etc.

Derucher, K.N. Heins, C.P. Bridge and pier protective systems and devices. Marcel Dekker. New York and Basle -----.

Tidal flows

Hiranandani, M.G. Chitale, S.V. Stream gauging. Manager of Publications. Delhi -----.

Siltation

Engineering Societies Monograph Committee. Estuary and coastline hydrodynamics. McGraw-Hill Book Co. Inc. New York -----.

Cofferdams

Lacroix, -----. Design, construction and performance of cellular cofferdams. ASCE Conference. 1976

Groynes

Summers, L. Fleming, C.A. Groynes in coastal engineering. CIRIA TN 111.

Lock gates

-----. Dock and Harbour Authority. June 1987

Dock walls

Greeves, I.S. London docks 1800–1980. Thomas Telford Ltd. London -----.

Corrosion

Dismuke, T.D. (ed). Handbook of corrosion protection for steel structures in marine environments. American Iron and Steel Institute, Washington USA 1981.

Historical background

Singer,---. Holmyard, ---. Hall, ---. Williams, ---. (Ed). History of technology. OUP 1985

Statutory, regulating and other authorities in the United Kingdom

British Ports Association,
Commonwealth House, 1–19, New Oxford Street, London WC1A 1DZ.
Corporation of Trinity House,
Trinity House, Tower Hill, London EC3N 4DN
The Company of Watermen and Lightermen of the River Thames,
16, St Mary-at-Hill, London EC3 8EE.
The Port of London Authority,
Tilbury Docks, Tilbury, Essex.
The Medway Ports Authority,
Sheerness Dockyard, Sheerness, Kent.
The Association of Water Authorities,
1, Queen Anne's Gate, London SW1.
Central Scotland Water Development Board,
Balmore, Torrance, Glasgow G64 4AJ.
Northern Ireland Water Service,
Department of the Environment for Northern Ireland,
Stourmont, Belfast BT4 3SS.
British Oil Spill Control Association,
33–38, Leman Street, London E1 8EW.
The Institute of Water and Environmental Management,
15, John Street, London WC1N 2EB.

Water Authorities in the United Kingdom

Anglian Water Authority,
Ambury Road, Huntingdon, Cambs. PE18 6NZ.
Northumbrian Water Authority,
Northumbria House, Regent Centre, Gosforth, Newcastle upon Tyne NE3 3PX.
North West Water Authority,
Dawson House, Great Sankey, Warrington WA5 3LW.
Severn-Trent Water Authority,
Abelson House, 2297 Coventry Road, Sheldon, Birmingham B26 3PU.

PRACTICAL TIDEWAY CONSTRUCTION

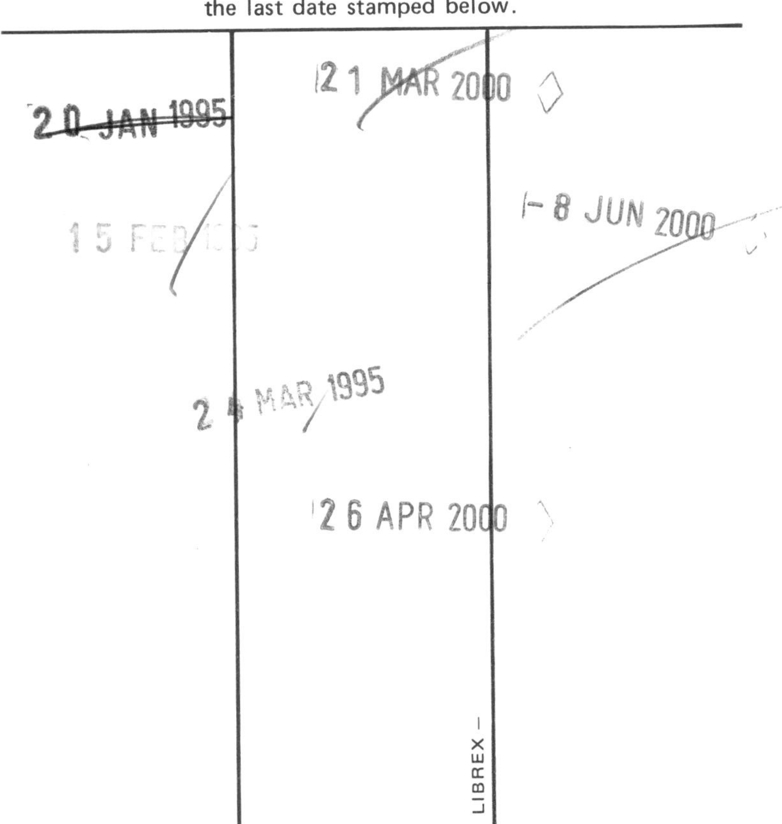

Final stage of mooring arrangements.
Placing the 30 metre long hinged brow.

HMS Belfast · London